T0140409

Terrorism, Security, and Computation

Series Editor

V. S. Subrahmanian
Department of Computer Science and Institute for Security, Technology and
Society, Dartmouth College, Hanover, NH, USA

The purpose of the Computation and International Security book series is to establish the state of the art and set the course for future research in computational approaches to international security. The scope of this series is broad and aims to look at computational research that addresses topics in counter-terrorism, counter-drug, transnational crime, homeland security, cyber-crime, public policy, international conflict, and stability of nations. Computational research areas that interact with these topics include (but are not restricted to) research in databases, machine learning, data mining, planning, artificial intelligence, operations research, mathematics, network analysis, social networks, computer vision, computer security, biometrics, forecasting, and statistical modeling. The series serves as a central source of reference for information and communications technology that addresses topics related to international security. The series aims to publish thorough and cohesive studies on specific topics in international security that have a computational and/or mathematical theme, as well as works that are larger in scope than survey articles and that will contain more detailed background information. The series also provides a single point of coverage of advanced and timely topics and a forum for topics that may not have reached a level of maturity to warrant a comprehensive textbook

More information about this series at http://www.springer.com/series/11955

V. S. Subrahmanian • Chiara Pulice
James F. Brown • Jacob Bonen-Clark

A Machine Learning Based Model of Boko Haram

Foreword by Geert Kuiper

V. S. Subrahmanian
Department of Computer Science
Dartmouth College
Hanover, NH, USA

Chiara Pulice
Department of Computer Science
Dartmouth College
Hanover, NH, USA

James F. Brown
Department of Computer Science
Dartmouth College
Hanover, NH, USA

Jacob Bonen-Clark
Institute for Advanced Computer Studies
University of Maryland, College Park
Maryland, MD, USA

ISSN 2197-8778   ISSN 2197-8786   (electronic)
Terrorism, Security, and Computation
ISBN 978-3-030-60616-9   ISBN 978-3-030-60614-5   (eBook)
https://doi.org/10.1007/978-3-030-60614-5

This Springer imprint is published by the registered company Springer Nature Switzerland AG
The registered company address is: Gewerbestrasse 11, 6330 Cham, Switzerland

Foreword

The establishment of facts is of utmost importance in intelligence gathering as it is in science and so should be in policy and politics too. The collection, analysis and exploitation of facts, enhanced by data science and machine learning, is an area where intelligence, science and policy have to cooperate to enhance the quality of assessment. This study enhances situational understanding and helps to discern the intentions and predict the activities of terrorist organizations like Boko Haram. The study shows which courses of action, by the military on the ground or policy decisions in the capital, are related in due time to the stepping up of kidnapping and sexual violence. The study is a great contribution to international crisis management.

Director of Strategy and Research,   Geert Kuiper
Netherlands Ministry of Defense
The Hague, The Netherlands

Former Deputy Director of the Netherlands
Defence Intelligence and Security Service
The Hague, The Netherlands

Preface

Sometime in 2015–2016, Rebecca Goolsby of the Office for Naval Research in Arlington, Virginia, approached the first author of this book with the idea of writing a book on Boko Haram. She had seen the first author's previous books on computational predictive models of Lashkar-e-Taiba (Springer 2012) and the Indian Mujahideen (Springer 2013) and felt that Boko Haram was an ever-increasing threat to global peace that needed a deeper scientific understanding than was available at the time – unfortunately, this is still true 5 years later. A pioneer in the field of computational modeling of human socio-cultural behavior, Dr. Goolsby felt that such a computational modeling effort would enhance efforts to reduce Boko Haram's deadly attacks. This book is a result of those early conversations, one that took far longer to write than expected.

We systematically gathered data on Boko Haram's behavior from 2009 to 2016 in order to learn not one but many predictive models of its behavior. Since January 2019, we have been generating predictions of Boko Haram attacks on the first of every month: the predictions specify what Boko Haram will likely do during the next 6 months. As an example, the predictions made on July 1, 2020 will predict different kinds of attacks that Boko Haram will carry out during the July–December 2020 period. This "real world" forecasting also enabled us to test the accuracy of our predictive models on real forecasts rather than on the usual cross validation methods alone (though those techniques were used in our internal testing models).

Because many traditional predictive models are hard to explain, we leveraged the concept of Temporal Probabilistic (TP) rules in order to provide a detailed understanding and explanation of why a prediction might be correct. TP rules also enable counter-terrorism and international security analysts to better understand Boko Haram's behaviors. We hope readers – including counter-terrorism analysts, law enforcement personnel, policy makers, and diplomats – will find this book to be of some value in their efforts.

We thank Rebecca Goolsby for her strong support for this work. John Tangney (initially at the US Air Force Office of Scientific Research and now at the Office of Naval Research) got the first author started in 2004/2005 on the first efforts to build computational models of terrorist groups. Purush Iyer at the US Army Research

Office and Liz Bowman at the US Army Research Lab have also consistently supported our work in this arena. Sachin Grover, who was a master's student at the University of Maryland, provided some assistance at the beginning of this project. We are grateful to Sabrina Jain and Mary-Versa Clemens-Sewall for carefully proof reading the manuscript. Last but not least, we are grateful to ONR grants N000141612739 and N00014-16-1-2918 for supporting parts of this work.

Hanover, NH, USA   V. S. Subrahmanian
Hanover, NH, USA   Chiara Pulice
Hanover, NH, USA   James F. Brown
Maryland, MD, USA   Jacob Bonen-Clark

June 8, 2020

Contents

Chapter 1
Introduction

On the morning of April 15, 2014, "tax" day in the United States, the world woke up to learn that 276 young schoolgirls had been abducted by terrorists in the small town of Chibok in Borno State (Mbah 2019). Located in the north-eastern corner of Nigeria, Borno State is not far from the border that nation shares with Chad, Niger, and Cameroon. The deadly raid was carried out by Boko Haram, a terrorist group that – until that time – was mainly well known to intelligence agencies, policy analysts, and military analysts, and less known by the general public. Despite years of attacks that included looting, kidnapping, arson, suicide bombings, bombings, assassinations, and sexual violence, it was not until the Chibok attack that Boko Haram was finally thrust into the public eye.

Boko Haram, also known today as the West African Province (WAP) or Islamic State's West Africa Province (ISWAP), had its roots in Maiduguri, the capital of Borno State. Split amongst multiple religions and ethnicities, the northern portion of Nigeria is largely poor and Islamic, while the southern portion of Nigeria is comparatively wealthier and more Christian. A deadly cocktail of factors have led to a host of tensions in Nigeria: religious divides between Christians and Muslims, tensions caused by the relative wealth of oil-rich Southern Nigeria and the relative poverty of the agricultural economy of Northern Nigeria, high levels of unemployment, a poor educational system, endemic corruption, a broadening social divide between the haves and the have nots and a lack of faith in the police, military, and justice system.

It was amidst this backdrop that Boko Haram was founded in 2002 in order to fight westernized education (the name "Boko Haram" is more or less synonymous with "Western education is forbidden"). Boko Haram is a fundamentalist Islamist group that aims to pull Northern Nigeria back to its Islamic roots and establish sharia law across Nigeria based on fundamentalist Wahhabi principles. Led by its founder Mohammed Yusuf, Boko Haram initially founded a religious school in Maiduguri (Ford 2014) which preached a fundamentalist brand of Islam and, over

time, became a breeding ground for jihadis, leading to tension between the group
and Nigerian security forces. These tensions boiled over into open conflict in July
2009 when Boko Haram clashed with security forces for the first time (Brun 2015).
Nigerian police quickly retaliated by capturing Mohammed Yusuf and executing
him in a public – and extrajudicial – show of force (Brun 2015).

Boko Haram's descent into an orgy of violence started shortly thereafter. After
regrouping under its new leader Abubaker Shekau, Boko Haram launched new
operations within a year. In the 10+ years that have passed since then, Boko Haram
fighters have carried out increasingly bold and devastating attacks including bombing the United Nations building in 2011 (Nossiter 2011), kidnapping several French
tourists from Cameroon in 2013 (Alsop 2013), capturing a military base in Baga in
2015 (BBC News 2015a, b), to a return to kidnapping over 100 schoolgirls in
Dapchi in 2018 (BBC News 2018a, b).

Many excellent books have already been written about Boko Haram (Thurston
2017; Matfess 2017; MacEachern 2018; Akaiso 2018). In contrast to these efforts,
this book is the first to analyze Boko Haram's actions using the type of predictive
models that advanced IT companies such as Google, Amazon, and Facebook use in
their products. It is the third in a series of books on computational models of terrorist groups – the first (Subrahmanian et al. 2012) focused on Lashkar-e-Taiba (LeT),
the group responsible for the Mumbai attacks of 2008, while the second
(Subrahmanian et al. 2013) focused on the Indian Mujahideen, a group that has been
closely allied with LeT.

Our book is based on a simple idea. Large IT firms such as Google, Amazon, and
Facebook have reduced billions of human beings each to one or more rows in a giant
spreadsheet or database and then used machine learning to learn predictive models
about the behavior of individuals. In the same way, data about terrorist groups can
also be reduced to a spreadsheet and analyzed using machine learning algorithms in
order to elicit behavioral models, generate forecasts of their behavior, and develop
methods to shape policy against them – all through the use of artificial intelligence
and machine learning techniques. In particular, the computational models used in
this book have been used to generate over one year of "monthly forecasts" of attacks
by Boko Haram from Jan 2019 to Feb 2020 before the publication of this book.

Table 1.1 below shows the accuracy of our forecasting models over this one year
"live" forecast window for different kinds of attacks. On the first of every month
during the above time frame, we put out a forecast specifying whether a given type
of attack (e.g. arson) would occur that month, sometime in the next 2 months, sometime in the next 3 months, and so forth. These correspond to the columns shown in
Table 1.1. For each type of attack, we computed four metrics: precision, recall,
F1-score and accuracy. Precision refers to the percentage of predictions made of
attacks that were actually correct. For instance, if we made 6 predictions that an
attack A would happen during the next month and the attack did happen in 4 of
those 6 predicted months, then the Precision in the 1-month column for attack A
would be 67%. Recall refers to the percentage of the true number of occurrences of
an attack that were in fact predicted. For instance, if an attack B happened in 5 of
the next 6 months and we predicted that attack B would happen in 4 of those months,

Table 1.1 Summary of real-world predictions made by our system. Predictions made on the first of each month from January 1 2019 to February 1 2020

Abduction						
Time period	1	2	3	4	5	6
Recall	90%	91%	100%	100%	100%	100%
Precision	90%	100%	100%	100%	100%	100%
Accuracy	82%	91%	100%	100%	100%	100%
F1	0.90	0.95	1.00	1.00	1.00	1.00

Arson						
Time period	1	2	3	4	5	6
Recall	91%	91%	100%	100%	100%	100%
Precision	100%	100%	100%	100%	100%	100%
Accuracy	91%	91%	100%	100%	100%	100%
F1	0.95	0.95	1.00	1.00	1.00	1.00

Attempted bombings						
Time period	1	2	3	4	5	6
Recall	0%	100%	89%	100%	100%	100%
Precision	0%	73%	80%	91%	100%	100%
Accuracy	27%	73%	73%	91%	100%	100%
F1	0.00	0.84	0.84	0.95	1.00	1.00

Bombings						
Time period	1	2	3	4	5	6
Recall	25%	44%	60%	90%	100%	100%
Precision	50%	80%	86%	100%	100%	100%
Accuracy	27%	45%	55%	91%	100%	100%
F1	0.33	0.57	0.71	0.95	1.00	1.00

Civilian casualties						
Time period	1	2	3	4	5	6
Recall	100%	100%	100%	100%	100%	100%
Precision	100%	100%	100%	100%	100%	100%
Accuracy	100%	100%	100%	100%	100%	100%
F1	1.00	1.00	1.00	1.00	1.00	1.00

Looting						
Time period	1	2	3	4	5	6
Recall	82%	100%	100%	100%	100%	100%
Precision	100%	100%	100%	100%	100%	100%
Accuracy	82%	100%	100%	100%	100%	100%
F1	0.90	1.00	1.00	1.00	1.00	1.00

Targeting of public sites						
Time period	1	2	3	4	5	6
Recall	43%	91%	73%	82%	91%	100%
Precision	100%	100%	100%	100%	100%	100%

(continued)

Table 1.1 (continued)

Targeting of public sites						
Accuracy	64%	91%	73%	82%	91%	100%
F1	0.60	0.95	0.84	0.90	0.95	1.00

Targeting of security installations						
Time period	1	2	3	4	5	6
Recall	82%	100%	100%	100%	100%	100%
Precision	100%	100%	100%	100%	100%	100%
Accuracy	82%	100%	100%	100%	100%	100%
F1	0.90	1.00	1.00	1.00	1.00	1.00

Targeting civilians for their beliefs						
Time period	1	2	3	4	5	6
Recall	20%	25%	38%	100%	100%	100%
Precision	100%	50%	50%	90%	100%	100%
Accuracy	64%	27%	27%	90%	100%	100%
F1	0.33	0.33	0.43	0.95	1.00	1.00

Targeting civilians indiscriminately						
Time period	1	2	3	4	5	6
Recall	100%	100%	100%	100%	100%	100%
Precision	100%	100%	100%	100%	100%	100%
Accuracy	100%	100%	100%	100%	100%	100%
F1	1.00	1.00	1.00	1.00	1.00	1.00

Sexual violence						
Time period	1	2	3	4	5	6
Recall	83%	100%	100%	100%	100%	100%
Precision	56%	73%	82%	91%	100%	100%
Accuracy	55%	73%	82%	91%	100%	100%
F1	0.67	0.84	0.90	0.95	1.00	1.00

Suicide bombings						
Time period	1	2	3	4	5	6
Recall	89%	90%	82%	100%	100%	100%
Precision	89%	100%	100%	100%	100%	100%
Accuracy	82%	91%	82%	100%	100%	100%
F1	0.89	0.95	0.90	1.00	1.00	1.00

then the Recall in the 1-month column for attack A would be 80%. The F1-score[1] is a metric that combines precision and recall: for the F1-score to be high, both precision and recall must be high. The "Accuracy" metric is the ratio of the total number of correct predictions made (attack will happen next month; attack will not happen

---

[1] Formally, $F1 = \dfrac{2 * P * R}{P + R}$ where P, R are precision and recall, respectively. It is easy to verify that the number assumes a low value if either precision or recall is low, leading to a small F1-score. Thus, a high F1-score can only be obtained when both P and R are high.

next month) to the total number of predictions made. In our case, we predict if an attack of type A will occur next month or not for each of the 14 months considered. Accuracy merely looks at how many of these are correct. So if our prediction agreed with the ground truth for 13 of the 14 months, then Accuracy would be 93%.

Accuracy is generally viewed as a highly flawed metric. We usually expect good models to have a high precision and recall – and hence a high F1-score.

Though many of the metrics shown in Table 1.1 are very high, it is important to also note what is not predicted. We do not (and currently cannot) predict *where* a given type of attack will take place – so for example, the high F1-scores for suicide bombings in Table 1.1 make high quality predictions, but make no statement about where the attack will occur. Moreover, the time frame is also not certain: the 100% precision and 90% recall when predicting suicide bombings 2 months ahead merely states whether suicide bombings will occur sometime in the next 2 months or not, rather than exactly when. As a consequence, these forecasts can be used to: (i) allocate intelligence assets to further investigate the group's behavior during a given time frame, (ii) take appropriate defensive measures such as moving more troops or security personnel to regions where security personnel believe a predicted attack type has a high chance of occurring, and (iii) alerting local officials via appropriate threat level notifications.

1.1 Organization of this Book

The rest of this book is organized as follows.

Chapter 2 contains a detailed description of the origins of Boko Haram. Starting with a brief description of Nigeria, it describes the origins of Boko Haram, its motivation and ideology, its geographic area of operations, it leadership, and the timeline from the start of the Boko Haram insurgency in 2009 to the current state of the insurgency. The chapter also provides broad coverage of actions taken by the Nigerian government, local and national security forces, as well as international actions such as asset freezes and trials of Boko Haram personnel. Simply put, it provides a brief history of Boko Haram.

Chapter 3 describes the temporal probabilistic (TP) rule formalism—this chapter can be skipped by those not interested in the technology. The chapter describes how these rules were automatically extracted from our Boko Haram dataset.

Chapter 4 looks at one class of attacks, sexual violence, carried out by Boko Haram. Boko Haram carried out sexual violence in 44 of the 89 months included in our study. The chapter identifies key factors linked to the occurrence of sexual violence by Boko Haram. The occurrence of sexual violence in a given month is linked to lack of explicit advocacy for religious rule in the preceding months, imprisonment of Boko Haram personnel in preceding months, and lack of explicit Boko Haram communications that are aimed at the Nigerian Government. The chapter presents several complex rules that are good predictors of sexual violence by Boko Haram.

Chapter 5 looks at another class of attacks, suicide bombings, which occur in almost half the months considered in our study. A favorite instrument of Boko Haram, suicide bombers often include women and young children. The chapter presents temporal probabilistic rules that are predictive both of months when Boko Haram does carry out suicide bombings as well as months where it does not do so. It also contains a discussion of the different types of factors linked to suicide bombings in subsequent months.

Chapter 6 describes abductions, yet another one of Boko Haram's devastating actions. Often focused on children who are then subjected to a litany of horrors, abductions occurred in approximately half of the months that we studied. This chapter looks both at abductions carried out by Boko Haram as well as when Boko Haram releases abducted persons. Abductions seem to be linked to factors such as Nigerian government actions that shut down Boko Haram locations/camps, no executions of Boko Haram personnel by Nigerian forces, and the nature of Boko Haram's communications. These, and other factors predictive of future abductions and prisoner releases by Boko Haram, are discussed in detail in Chap. 6.

Chapter 7 focuses on arson, yet another major instrument of terror used by Boko Haram. This chapter focuses on identifying factors that are predictive of months where arson attacks are carried out as well as months where Boko Haram does not carry out any arson attacks.

Chapter 8 focuses on factors that predict four types of attacks: the targeting of government officials, looting, attempted bombings, and the targeting of security installations. This chapter discusses temporal probabilistic rules that are capable of predicting the first three types of attacks. For the fourth type of attack, namely the targeting of security installations, this chapter presents TP rules that predict the absence of such attacks in a given month.

Chapter 9 describes the policies we derived in order to mitigate attacks carried out by Boko Haram. We start out with a description of the methodology used to derive these policies (though this can be skipped by a reader not interested in technical details). We then describe five policies we have derived from the data that may help reduce the different types of violence carried out by Boko Haram.

A set of appendices provides rich supplementary material. Specifically, Appendix A is a list of all terrorist attacks carried out by Boko Haram and its predecessors during the 2009–2019 period. Appendix B is a list of all TP-rules presented in this book. Appendix C describes how the data used in this book was collected and coded. It will examine all the types of variables for which we gathered data. Appendix D lists the data we used about the major factors found to be linked to Boko Haram attacks. Appendix E shows the prediction reports we generated from January 12, 2019 to Feb 2020. These prediction reports are updated to assess the accuracy of the predictions.

1.2 How to Read This Book

Readers with no interest in the technology used to derive the results in this book can safely skip Chap. 3 without losing any insights into Boko Haram's behavior. Readers with an interest in the technology will find technical details provided in these portions of the book; we do not provide the detailed mathematics used to derive our algorithms here but explain them in a simple manner so as to convey the basic ideas of how these computations are done. The details are provided in the technical references provided in these chapters.

1.3 Summary Statistics About Boko Haram's Violent Activities

Though the origins of Boko Haram go back to 2002, our study derives temporal probabilistic (TP) rules using data from the 2009–2016 time period. Because we have been putting out "live" monthly forecasts on the first of every month since January 1st, 2019 (i.e. for over a year at the time of writing this chapter), we will also cover attack data below that is valid until the end of 2019.

Table 1.2 below shows broad statistics about the different kinds of attacks we have captured in our data.

Table 1.2 Different types of attacks carried out by Boko Haram, 2009–2019

Attacks	2009[a]	2010	2011	2012	2013	2014	2015	2016	2017	2018	2019
Abductions	0	0	0	0	2	10	7	8	9	8	11
Arson	1	0	1	5	8	12	9	9	7	10	12
Assassinations	0	1	3	2	0	0	0	0	1	0	0
Attack Schools	0	0	1	3	3	3	2	0	4	1	2
Attempted Bombings	0	0	3	8	0	1	7	10	9	4	4
Bombings	0	1	8	7	2	8	8	6	2	5	5
Casualties - Civilian	1	3	10	11	10	12	12	11	12	12	12
Casualties - Security Forces	1	5	10	11	11	10	12	10	12	12	12
Looting	0	0	0	0	1	6	4	4	8	11	12
Releases	1	0	0	0	1	6	7	10	2	2	2
Sexual Violence	0	0	0	0	9	12	12	11	11	10	6
Suicide Bombings	0	0	2	7	0	7	11	11	12	9	10
Targ Civ for Beliefs	0	2	7	7	4	6	8	4	4	3	5
Targ Civ Indiscriminate	1	2	9	10	10	12	12	12	12	10	12
Targ Gov Officials	0	2	6	9	3	3	4	1	1	3	2
Targ Public Site	1	0	6	3	0	6	1	1	6	8	8
Targ Public Transport	0	0	1	0	0	4	8	0	2	1	2
Targ Security Install	1	3	6	10	8	8	9	2	8	9	12

[a]Records start during July 2009

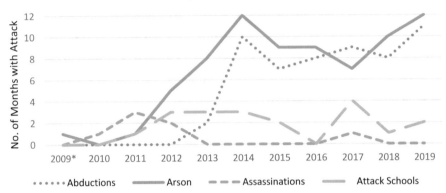

Fig. 1.1 Number of months (2009–2019) when Boko haram carried out abductions, arson attacks, assassinations, and attempted bombings

We can drill further down into each attack type. For example, Fig. 1.1 shows the number of months in the years 2009–2019 which experienced abductions, arson attacks, assassinations, and attacks on schools. While these numbers were relatively small in 2009, arson attacks and abductions took off after 2009. Attacks on schools would happen from time to time, while assassinations were relatively rare.

Figure 1.2 shows an analogous figure in the case of incidents of attempted bombings, bombings, civilian casualties, and security forces casualties. We see that since 2011, Boko Haram has consistently caused the deaths of civilians and security personnel. Non suicide bombings became more and more frequent until they fell off in 2013. The monthly use of non-suicide bombings would peak in 2016. The monthly occurrence of attempted bombings is somewhat similar to that of non-suicide bombings.

Figure 1.3 shows an analogous figure that shows the number of months in each year that saw incidents of lootings, hostage releases, sexual violence and suicide bombings. While suicide bombings were not very frequent prior to 2014, the years since 2014 have seen them happen in at least in 8 months per year. In fact, from 2015 to 2019, suicide bombings occurred in 9–12 months of every year. Sexual violence has been nearly constant since 2013. As of late 2018, this type of violence has waned slightly, but is still occurring one out of every two months. The number of months per year with looting has been steadily rising over time since 2013. During 2019 an incident of looting was reported every month. Hostage releases were first observed in 2013 and reached the peak in 2016; since then, hostage releases have been infrequent.

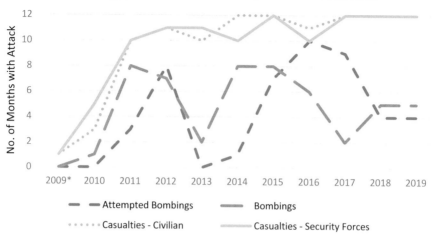

Fig. 1.2 Number of months (2009–2019) when Boko haram carried out looting, releases of abducted persons and sexual violence

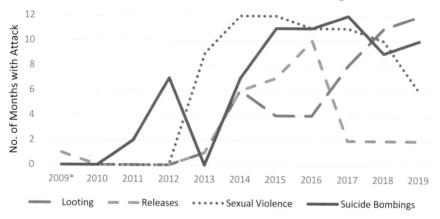

Fig. 1.3 Number of months (2009–2019) when Boko haram carried out looting, hostage releases, sexual violence, and suicide bombings

Figure 1.4 depicts an analogous figure that shows the number of months in each year that saw incidents targeting public sites, public transportation, and security installations. The targeting of public transportation has been infrequent except during 2015. On the other hand, targeting of security installations (which are relatively

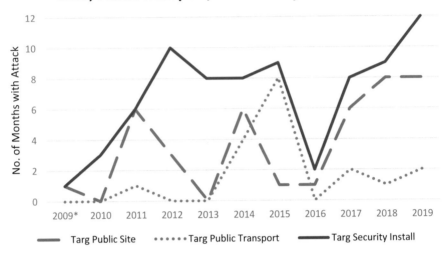

Fig. 1.4 Number of months (2009–2019) when Boko haram targeted public sites, public transportation, and security installations

hard targets) was relatively infrequent during the beginning of the insurgency (2009–2011) but has increased steeply from 2014 onward to the point that between 2015 and 2019, we have seen such attacks happen during 8–12 months of every year. The targeting of public sites has waxed and waned over time, but during 2017, 2018, and 2019 these types attacks became a consistent part of Boko Haram's arsenal.

Figure 1.5 presents an analogous figure that shows the number of months in each year that saw incidents targeting civilians for their beliefs, targeting civilians indiscriminately, and targeting government officials. The indiscriminate targeting of civilians has been constant since 2011 with most months having an incident of this type. Boko Haram's targeting of government officials has been less consistent. The peak of this type of targeting came in 2011 and 2012 when more than six months out of each year saw these types of incidents. Since 2012, the targeting of government officials has been less frequent. Finally, there is Boko Haram's targeting of civilians for their beliefs. Targeting of this type has been occurring throughout the conflict, and the number of months with such incidents peaked in 2015. During this year it was reported that this type of target occurred during eight months. Since 2015, this type of target has become more infrequent.

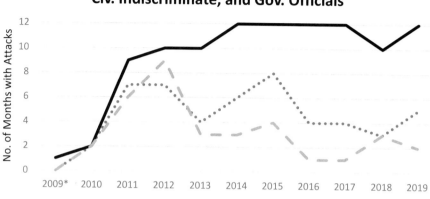

Fig. 1.5 Number of months (2009–2019) when Boko haram targeted civilains for their beliefs, civilians indiscriminately, and government officials

1.4 Summary of Significant Temporal Probabilistic (TP) Rules

This book present 39 TP-rules that are capable of explaining the conditions under which 9 types of actions are taken by Boko Haram. These actions include:

- Sexual violence
- Suicide bombings
- Attempted bombings
- Arson
- Looting
- Targeting Security Officials
- Targeting Security Installations
- Abductions
- Release of Abducted Persons

Note that not all actions that we can predict well have easy to understand explanations – this is a problem with Machine Learning that many researchers are currently working on (Subrahmanian and Kumar 2017). For instance, the attack "Targeting of Public Sites" shown in Table 1.1 is one that we can predict well, but for which a human-intelligible (as opposed to machine intelligible) explanation is harder to come by.

Table 1.3 below summarizes the relationship between certain conditions prevailing in Boko Haram's operating environment and the occurrence or non-occurrence of these events. The presence of a "+" indicates a positive link, while the presence of a "−" indicates a negative link between the condition and the action.

Table 1.3 Relationship between key variables and different types of actions taken by Boko Haram

	Sex. Violence	Suicide Bombings	Att. Bombings	Arson	Targ Gov Officials	Targ Security Install	Abductions	Abduction: Releases	Looting
Gov* Arrests or Imprisons BH	+	+							
BH has Foreign Members	-	-							
Comms Address Gov*	-	-							
BH Declines Negotiations	-	+							
Gov* Shuts Down BH Locations	+	+		+	+		+		
BH using Child Soldiers	+	+	+	+			+	+	
Military Aid to Gov* Suspended	-			-		+	-		+
BH Members on Trial		-						-	
BH Supports Non-State Actor Groups		+				-	-		
Gov* Executes BH People				-	-		-		
BH Designated Intl. Terror Org					+				
Gov* Has Closed Borders									-
BH Assets Frozen				+					
Forcible Resettlement by Gov*					-				

NSAG **Non-State Armed Group,** *Gov* **Government of Nigeria**

For instance, the two "+"s in the row on Arrests/Imprisonment of BH personnel say that there is a positive link between these variables. In particular, this says that we were able to derive TP-rules that suggest that when the Government of Nigeria arrests Boko Haram personnel and/or imprisons them, we can expect to see sexual violence and suicide bombings sometime in the next few months. On the other hand, the "−" in the last row suggests that targeting of government officials by Boko Haram is inversely linked to forcible resettlement of civilians by the Government of Nigeria.

In total, our study identified over 15 factors that are linked to Boko Haram's activities – including both the occurrence and non-occurrence of various types of attacks. Of these, the most significant factors are the following: Shutdown of Boko Haram locations by the Government of Nigeria, Executions of Boko Haram personnel by the Government of Nigeria, Use of Child Soldiers by Boko Haram, Suspension of Military Aid to the Government of Nigeria. We discuss each of these below.

Shutdown of Boko Haram Locations The Government of Nigeria's security forces frequently take actions to shut down Boko Haram's offices and other locations. Our results show that such shutdowns have mixed effects. For instance, not shutting down Boko Haram locations is positively linked to non-occurrence of sexual violence by Boko Haram in subsequent months. In fact, the same is true w.r.t. suicide bombings, i.e. not shutting down Boko Haram locations is positively linked to non-occurrence of suicide bombings in subsequent months. The same is also true of arson attacks – not shutting down Boko Haram locations is linked to the non-occurrence of arson attacks in subsequent months.

When Boko Haram locations are shutdown by the Government of Nigeria, we do see that two types of attacks – abductions and targeting of government officials – do occur a few months later. These results suggest that shutting down Boko Haram locations may be counter-productive in terms of reducing the subsequent occurrence of sexual violence, suicide bombings, arson, and abductions.

Executions of Boko Haram Personnel by the Government of Nigeria The Government of Nigeria's security forces have faced numerous accusations of impropriety over the years (Boghani 2014). Our work finds that reports of executions of Boko Haram personnel by Nigerian security forces are negatively linked with incidents of arson, abductions, and targeting of government officials. For instance, abductions tend to occur in months following those months where Boko Haram personnel were not (reportedly) executed. Likewise, arson attacks tend to occur in months after ones where Nigerian security forces did not execute Boko Haram personnel. The same pattern holds in terms of explicit targeting of government personnel by Boko Haram – when Boko Haram's personnel are not executed, we see that they target government officials shortly thereafter. These results suggest that the executions of Boko Haram personnel serve as a strong deterrent to Boko Haram – when such executions do not occur, Boko Haram carries out significant attacks in succeeding months. While we strongly oppose the idea of governments carrying out extrajudicial executions, our results show that a "big stick" approach seems to reduce some terrorist acts by Boko Haram.

Suspension of Military Aid to the Government of Nigeria The Government of Nigeria gets significant military aid from foreign nations. As an example, Figs. 1.6 and 1.7 below show the amounts of security assistance released by the US Department of Defense and the US Department of Homeland Security respectively during the period 2009–2020.

Our results show a negative link between suspension of such aid to the Government of Nigeria and arson and targeting of security installations, but there is a positive link between such suspensions and looting. Suspension of military aid to

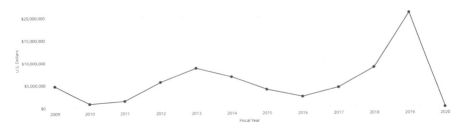

Fig. 1.6 US Department of Defense Aid to Nigeria during the 2009–2020 Time Frame. (Source: https://www.securityassistance.org/data/program/military/Nigeria/2009/2020/Defense%20 Department/Global//)

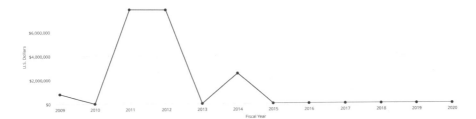

Fig. 1.7 US Department of Homeland Security Aid to Nigeria during the 2009–2020 Time Frame. (Source: https://www.securityassistance.org/data/program/military/Nigeria/2009/2020/ Homeland%20Security/Global//)

Nigeria is linked to a reduction in incidents of both arson and attacks on security installations in subsequent months. However, just the opposite occurs as far as looting is concerned – suspension of military aid is followed by periods when Boko Haram goes on looting sprees.

Reports of Use of Child Soldiers by Boko Haram Child soldiers are heavily used by Boko Haram (Bukarti 2019). Our results discover a strong positive relationship between incidents of virtually every kind of attack carried out by Boko Haram and reports of the use of child soldiers by Boko Haram. In particular, we note that when such reports are present, there is a positive link to occurrence of sexual violence, suicide bombings, attempted bombings, arson, and abductions. Release of abducted prisoners by Boko Haram is also positively linked to reports of the use of child soldiers.

1.5   Summary of Policy Recommendations

Our team leveraged methods for generating policies building on top of the ideas and algorithms proposed in (Simari et al. 2013) – a simplified version is described in Chap. 3. In all, the policy analytics algorithm discovered 5 policies that we call P1-P5 which are shown in Table 1.4 below.

All the policies (P1)–(P5) proceed under the assumption that Boko Haram will continue to advocate for the imposition of Islamic law in Nigeria. This does not seem to be an unreasonable assumption as Boko Haram has consistently parroted this line since its very inception.

Figure 1.8 shows the situation – in particular, it shows which actions are common to all the policies and how the policies vary from one another.

Foreign Military Aid   All the policies require the delivery of foreign military aid to the Nigerian government. Table 1.5 below shows the level of US foreign military aid to Nigeria.

Foreign military aid to Nigeria has been controversial in the US – a mix of allegations of rampant corruption (UNODC 2019) and misdeeds by Nigerian security forces (Boghani 2014) have severely affected US economic and military aid to Nigeria. The same is true of economic and military aid from the UK (Wintour 2019). Nonetheless, all 5 of our policies agree on one thing, namely that foreign aid to the Nigerian military needs to be ramped up if the US wishes to reduce attacks carried out by Boko Haram.

Clampdown on Recruitment, Training & Deployment of Child Soldiers   Child soldiers help fuel Boko Haram's deadly attacks. For example, Boko Haram pioneered the use of children as suicide bombers. Young females are often used for their ability to discreetly infiltrate public areas in order to bomb them (Parkinson and Hinshaw 2019). Our models show that clamping down on Boko Haram's ability to transform children into a lethal fighting force will significantly help deter future attacks by Boko Haram. Simply put, the Government of Nigeria needs to invest more resources and develop a more coherent strategy to protect Nigerian children.

Imprisonment of Boko Haram Personnel   Counter-terrorism actions which seem just and logical do not always have the intended effects. Our models show that months in which the Government of Nigeria successfully imprisons Boko Haram personnel are often followed by various kinds of attacks such as suicide bombings, looting, arsons, and abductions. As a consequence, our models make the counter-intuitive suggestion that Boko Haram personnel not be imprisoned. This does not mean that those engaged in planning or executing attacks not be incarcerated. Perhaps high profile arrest campaigns should not be carried out and those arrests that are made should not be the subject of high profile publicity campaigns.

Table 1.4 Summary of the assumptions and actions of different stakeholders in order to reduce attacks by Boko haram

Policy	Actions to be taken in policy
P1	**Assumption**: Boko Haram continues to advocate for Islamic religious rule
	Actions to be taken by the Government of Nigeria
	1. The Nigerian Government does not ramp up steps to imprison members of Boko Haram
	2. The Nigerian Government does not close down Boko Haram locations
	3. The Nigerian Government clamps down on Boko Haram's recruitment, training, and deployment of child soldiers
	4. Nigerian security forces execute "members" of Boko Haram
	Actions to be taken by other parties
	1. Foreign nations/ organizations provide military aid to the Government of Nigeria
P2	**Assumption**: Boko Haram continues to advocate for Islamic religious rule
	Actions to be taken by the government of Nigeria
	1. The Nigerian government does not ramp up steps to imprison members of Boko Haram
	2. The Nigerian Government does not close down Boko Haram locations
	3. The Nigerian Government clamps down on Boko Haram's recruitment, training, and deployment of child soldiers
	4. The Nigerian Government forcibly resettles civilians in relevant areas
	Actions to be taken by other parties
	1. Foreign nations/ organizations provide military aid to the Government of Nigeria
P3	**Assumption**: Boko Haram continues to advocate for Islamic religious rule
	Actions to be taken by the Government of Nigeria
	1. The Nigerian Government does not ramp up steps to imprison members of Boko Haram
	1. The Nigerian Government does not close down Boko Haram locations
	2. The Nigerian Government clamps down on Boko Haram's recruitment, training, and deployment of child soldiers
	3. The Nigerian Government holds elections as scheduled
	Actions to be taken by other parties
	1. Foreign nations/ organizations provide military aid to the Government of Nigeria
P4	**Assumption**: Boko Haram continues to advocate for Islamic religious rule
	Actions to be taken by the Government of Nigeria
	1. The Nigerian Government does not ramp up steps to imprison members of Boko Haram
	2. The Nigerian Government clamps down on Boko Haram's recruitment, training, and deployment of child soldiers
	3. Nigerian security forces execute "members" of Boko Haram
	4. The Nigerian Government holds elections as scheduled
	Actions to be taken by other parties
	1. Foreign nations/ organizations provide military aid to the Government of Nigeria

(continued)

Table 1.4 (continued)

Policy	Actions to be taken in policy
P5	**Assumption**: Boko Haram continues to advocate for Islamic religious rule
	Actions to be taken by the Government of Nigeria
	1. The Nigerian Government does not ramp up steps to imprison members of Boko Haram
	2. The Nigerian Government clamps down on Boko Haram's recruitment, training, and deployment of child soldiers
	3. The Nigerian Government forcibly resettles civilians in relevant areas
	4. The Nigerian Government holds elections as scheduled.
	Actions to be taken by other parties
	1. Foreign nations/organizations provide military aid to the Government of Nigeria

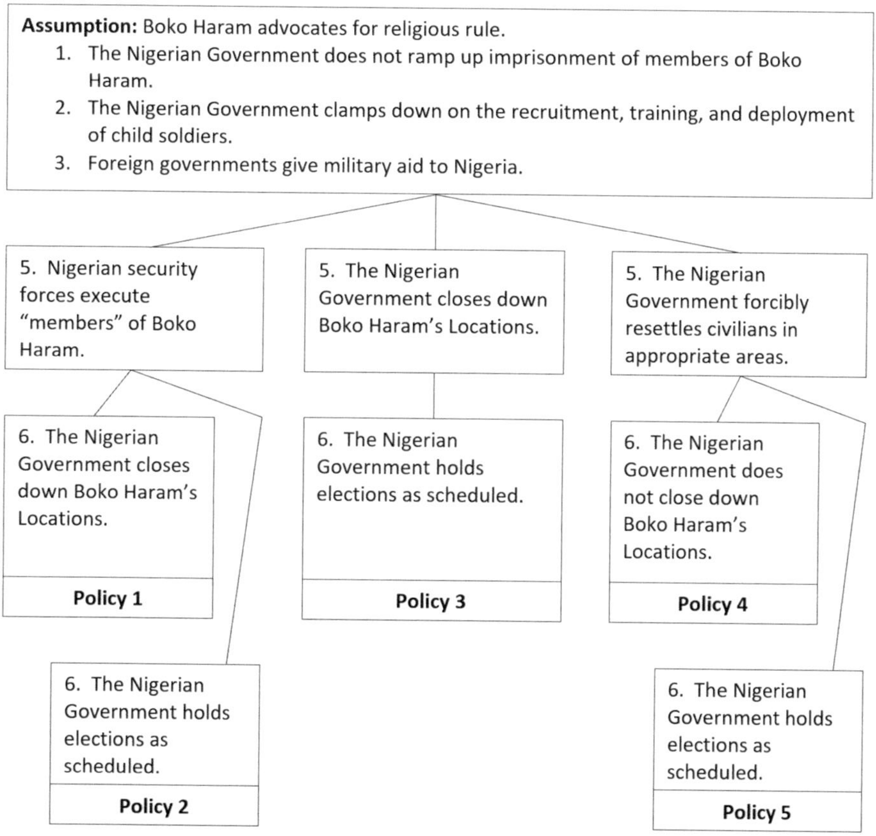

Fig. 1.8 Summary of policy actions and how they differ

Table 1.5 Summary of
U.S. Foreign aid to Nigeria

Year	Military aid millions (USD)	Economic aid millions (USD)
2011	11.53	629.77
2012	23.37	645.03
2013	10.86	698.12
2014	11.15	700.81
2015	15.62	645.84
2016	3.79	473.41
2017	12.64	584.18
2018	15.87	413.3
2019	20.91	345.81

Source: https://www.securityassistance.org/
nigeria (retrieved March 16 2020)

The above three types of actions are common to all the 5 policies derived by our system to mitigate Boko Haram's attacks. However, these policies also involve other types of actions.

Policy P1 suggests two additional actions: one, which is reasonably uncontroversial, is that the Government of Nigeria should shut down Boko Haram's locations such as offices. We do know that shutting down Boko Haram's locations will lead to a likely mitigation of the following types of attacks: sexual violence, suicide bombings, arson, and abductions.. However, a more controversial recommendation is that the Government of Nigeria should forcibly execute Boko Haram personnel – taken in conjunction with the recommendation that they not imprison Boko Haram personnel, this suggests that arrested Boko Haram personnel should just be executed. This is obviously morally repugnant and also a violation of international human rights norms, and so we do *not* recommend policy P1 but merely discuss it in passing as our algorithms do suggest these as possible policy options.

Policy P2 likewise suggests the morally unacceptable suggestion that the Government of Nigeria should carry out executions of Boko Haram personnel and includes the recommendation that the Government of Nigeria should conduct elections on schedule. As in the preceding case, we believe that this policy is not acceptable and is inconsistent with international human rights norms – hence, we do not recommend this option either.

Policy P3 is the most benign of all the suggested policies. In addition to the three actions common to all the five policies we computed, this policy suggests two additional actions. The first is that the Government of Nigeria should shut down Boko Haram's locations such as offices. The second is that the Government of Nigeria should hold free and fair elections on schedule. We strongly recommend this policy.

Policy P4 suggests two actions in addition to the 3 actions common to all five policies computed by our framework. The first is that the Government of Nigeria forcibly resettle civilians in certain areas – these would include areas such as the northeastern states on Nigeria and the areas near Lake Chad. These locations have been the hardest hit by the Boko Haram insurgency. The second is that the

Government of Nigeria close down Boko Haram locations such as offices. While this policy is not as abhorrent as policies P1 and P2, it does involve a difficult task of somewhat dubious morality – that of forcibly resettling civilians from areas such as Lake Chad and the northeastern states of Nigeria. We therefore find this option less preferable to policy P3 and do not recommend it.

Policy P5 suggests two actions in addition to the 3 actions common to all five policies computed by our framework. As in the case of Policy P4, it suggests that the Government of Nigeria forcibly resettle civilians in certain areas. The second action is to hold free and fair elections. Again, we find this option less preferable to policy P3 and do not recommend it.

In short, policy P3 emerges as the most feasible policy for the Government of Nigeria. We discuss this policy in more detail in Sect. 1.5.

1.6   The Most Feasible Policy Recommended to Combat Boko Haram

The most feasible policy to rein in Boko Haram appears to be Policy P3. Policy P3 contains both dos and don'ts as shown below.

The policy has two don'ts:

1. The Nigerian Government should not ramp up steps to imprison members of Boko Haram.
2. The Nigerian Government should not close down Boko Haram locations.

Both of these actions are offensive actions by the Government of Nigeria that Boko Haram may feel the need to respond to. For example, in one incident in June 2014, the Government of Nigeria arrested a leader of Boko Haram (Igboerotoenwo and Hemba 2014). Then, in September 2014, Boko Haram abducted two Nigerian women (Sieff 2016). During this raid, many men were killed while the women were thrown into a world of sexual slavery and forced marriages (Sieff 2016). On the other hand, the Government of Nigeria did not shut down any of Boko Haram's offices in April 2013. No arson attacks occurred in subsequent months.

We do not intend to suggest that any one of the four conditions listed in Policy (P3) by themselves alone imply the non-occurrence of subsequent attacks. **Rather, it is when all five of the actions in Policy (P3), four by the Government of Nigeria and one by international donors are performed concurrently, that our model suggests that attacks by Boko Haram can be reduced.**

In addition, Policy P3 requires that the Government of Nigeria takes two explicit actions.

1. The Nigerian Government clamps down on Boko Haram's recruitment, training, and deployment of child soldiers.
2. The Nigerian Government holds elections as scheduled.

The Government of Nigeria must clamp down on the use of child soldiers by Boko Haram (Bukarti 2019) as such soldiers are used in large numbers by Boko Haram. In addition to the morality of the matter and adherence to UN Security Council Resolution 1261 (adopted in Aug 1999), our model suggests that recruitment of child soldiers makes Boko Haram more lethal.

Elections are another hot button issue and their cancellation can lead to frustration and a feeling of disenfranchisement, while holding elections as planned have the opposite effect. In March 2015 for example, national elections for president were held and gave many in the embattled north a reason to celebrate (BBC News 2015a, b). For the next three months, April, May, and June, no abductions were reported. In addition, during April there were no reported arson attacks, and civilians were not reported to have been targeted for their beliefs.

Finally, as discussed earlier, we recommend that foreign nations/organizations provide a consistent stream of military aid to the Government of Nigeria. As shown in Table 1.4, the amount of military aid from the US to the Government of Nigeria has fluctuated from just \$3.79 M in 2016 to \$23.37 M in 2012. This represents a huge fluctuation in the volume of military aid, making it hard for Nigerian planners to consistently plan campaigns against Boko Haram.

1.7 Conclusion

With over 125 attacks in 2019 alone, Boko Haram is one of the deadliest terrorist groups in the world, spanning the nations of Nigeria, Cameroon, Chad and Niger.

This chapter discusses how advanced AI techniques can be used to predict various kinds of attacks carried out by Boko Haram including a discussion of 14 months of predictions made by our team since January 2019. We discuss the following key factors that are linked to attacks by Boko Haram: arrests of Boko Haram personnel, shutdown of Boko Haram locations (e.g. offices) by the Government of Nigeria, asset freezes, designation of Boko Haram as a terrorist organization, use of child soldiers by Boko Haram, suspension of military aid to the Government of Nigeria. We also discuss factors that are linked to the absence of different types of attacks by Boko Haram.

Finally, we identify and discuss a set of 5 policies that were derived from the data. All of these policies require that: (i) the Nigerian government does not ramp up imprisonment of Boko Haram's members, and (ii) the Nigerian Government clamps down on the recruitment, training, and deployment of child solders, and (iii) that foreign governments give military aid to the Nigerian government. Of the five policies, we recommend one policy P3 which, in addition to the 3 actions listed above, require the Nigerian Government to additionally perform the following actions:(iv) shut down Boko Haram locations, and (v) hold elections on schedule in a responsible and fair manner.

References

Akaiso D (2018) Misunderstanding Boko Haram: understanding the historic and ethnic causes of Nigeria's fundamentalist terrorist group. Soyounique Experts. Isbn: 978-1554838677

Alsop H (2013) France condemns Boko Haram kidnap video as 'cruelty' without bounds. The Telegraph. https://www.telegraph.co.uk/news/worldnews/europe/france/9893656/France-condemns-Boko-Haram-kidnap-video-as-cruelty-without-bounds.html. Accessed 18 Mar 2020

BBC News (2015a) Boko Haram CRISIS: Nigeria's Baga town hit by new assault. https://www.bbc.com/news/world-africa-30728158. Accessed 28 July 2019

BBC News (2015b) Nigeria election: Muhammadu Buhari wins presidency. https://www.bbc.com/news/world-africa-32139858. Accessed 19 Mar 2020

BBC News (2018a) Chibok girls: many abductees dead, says journalist. https://www.bbc.com/news/world-africa-43767490. Accessed 28 Nov 2018

BBC News (2018b) Dapchi girls: freed Nigerian girls tell of kidnap ordeal. https://www.bbc.com/news/43489217. Accessed 4 Aug 2019

Boghani P (2014) uncovering atrocities committed by Nigerian Security Forces. Frontline. https://www.pbs.org/wgbh/frontline/article/uncovering-atrocities-committed-by-nigerian-security-forces/. Accessed 19 Mar 2020

Brun H (2015) Boko haram: the early years. The Interpreter. https://www.lowyinstitute.org/the-interpreter/boko-haram-early-years. Accessed 11 July 2019

Bukarti AB (2019) Nigeria's child veterans are still living a nightmare, Foreign Policy, August 15 2019, https://foreignpolicy.com/2019/08/15/children-boko-haram-nigeria-borno-cjtf/. Retrieved March 16 2020

Ford J (2014) The origins of Boko Haram. The National Interest. https://nationalinterest.org/feature/the-origins-boko-haram-10609. Accessed 10 July 7 2019

Igboerotoenwo A, Hemba J (2014) Nigerian military arrest senior Boko Haram member: Police. Reuters. https://www.reuters.com/article/us-nigeria-blast/nigerian-military-arrest-senior-boko-haram-member-police-idUSKBN0ET2OD20140618. Accessed 19 Mar 2020

MacEachern S (2018) Searching for Boko Haram: a history of violence in Central Africa. Oxford University Press. Isbn: 978-0190492526

Matfess H (2017) Women and the War on Boko Haram: wives, weapons, witnesses. Zed Books. Isbn: 978-0691172248

Mbah F (2019) Nigeria's Chibok schoolgirls: five years on, 112 still missing. Al Jazeera. https://www.aljazeera.com/news/2019/04/nigeria-chibok-school-girls-years-112-missing-190413192517739.html. Accessed 23 July 2019

Nossiter A (2011) Islamic group says it was behind fatal Nigeria attack. The New York Times. https://www.nytimes.com/2011/08/29/world/africa/29nigeria.html Accessed 21 July 2019

Parkinson J, Hinshaw D (2019) 'Please, save my life.' A bomb specialist defuses explosives strapped to children. The Wall Street Journal. https://www.wsj.com/articles/please-save-my-life-a-bomb-specialist-defuses-explosives-strapped-to-children-11564156722. Accessed 19 Mar 2020

Sieff K (2016) They were freed from Boko Haram's rape camps. But their nightmare isn't over. The Washington Post. https://www.washingtonpost.com/world/africa/they-were-freed-from-boko-harams-rape-camps-but-their-nightmare-isnt-over/2016/04/03/dbf2aab0-e54f-11e5-a9ce-681055c7a05f_story.html. Accessed 6 Dec 2018

Simari G, Dickerson J, Sliva A, Subrahmanian VS (2013) Parallel abductive query answering in probabilistic logic programs. ACM Trans Comput Log 14(2):12

Subrahmanian VS, Kumar S (2017) Predicting human behavior: the next frontiers. Science 355(6324):489

Subrahmanian VS, Mannes A, Sliva A, Shakarian J, Dickerson J (2012) Computational analysis of terrorist groups: Lashkar-e-Taiba. Springer

Subrahmanian VS, Mannes A, Roul A, Raghavan RK (2013) Indian Mujahideen: computational analysis and public policy. Springer, Cham

Thurston A (2017) Boko Haram: the history of an African Jihadist movement. Princeton University Press. Isbn: 978-1786991461

UNODC (2019) Corruption in Nigeria: patterns and trends. https://www.unodc.org/documents/nigeria/Corruption_Survey_2019.pdf. Accessed 19 Mar 2020

Wintour P (2019) UK could boost military support to help Nigeria defeat Boko Haram. The Guardian.   https://www.theguardian.com/world/2019/may/01/uk-could-boost-military-support-to-help-nigeria-defeat-boko-haram. Accessed 19 Mar 2020

Chapter 2
History of Boko Haram

Nigeria is the most populous country in Africa, and it is also the largest economy in Africa. As of early 2019, Nigeria's Gross Domestic Product (GDP) was $376.284 billion; much of its GDP is generated by the nation's large oil industry (Oyekunle 2019). While Nigeria has the largest economy on the continent, poverty remains a huge issue. It is estimated that about half of the country lives in poverty (Mwai and Goodman 2019). In addition, the country's poverty issue is exacerbated by a north/south divide that splits the country. The southern portion of Nigeria has more wealth than the north, while having a smaller population. This imbalance of wealth stems from the south's control over the country's oil reserves (Campbell 2011). The north/south divide also occurs along religious lines with the south comprising mostly of Christians and the north mostly of Muslims. These differences contribute to the north's distrust of the south (Campbell 2011). While Nigeria's issues with poverty are significant, the country's modern history contains substantial adversity as well.

Since its independence from the United Kingdom in 1960, Nigeria has faced a series of hardships. From 1960 to 1998 Nigerian citizens lived under military rule (World Factbook 2019). During this period, the first of many coups occurred in 1966 and devastating civil war followed. Nigeria was engulfed in civil war from 1967 to 1970 and an estimated one million people were killed in the fighting (The Commonwealth 2019). From the end of the civil war to 1999, the country was run by several military governments. The first legitimate election took place in December of 1998 and a new constitution was approved in 1999 (The Commonwealth 2019). In this new constitution, a provision was made that permitted the practice of Sharia law, a code of laws derived from the teachings and values of Islam.

During the late 1990s and early 2000s, northern Nigeria saw many states implement versions of Sharia law that began to cover criminal law (Tertsakian 2004). Before this period, Sharia law had been used in practice, but it mainly covered civil law. The legal process and punishments used by these courts were often brutal and abusive. During the early 2000s, two women were sentenced to be stoned to death after being convicted of adultery (Tertsakian 2004). Other punishments handed out by courts included amputation as well as flogging (Tertsakian 2004). These aggressive applications of Islamic law demonstrate the commitment in the region to fundamentalist Islam. Together, fundamental religious fervor and dire economic conditions would help fuel one of the most dangerous terrorist groups of the twenty-first century.

2.1   Emergence and History of Boko Haram

2.1.1   Boko Haram's Beginnings

Boko Haram's origins can be found in the rise of fundamental Islam in Nigeria during the second half of the twentieth century. This religious fundamentalism resulted in the establishment of Sharia law in some of the northern states of Nigeria, and in this environment Boko Haram was founded (Ford 2014). Boko Haram was founded in 2002 in Maiduguri, the capital of the Nigerian state of Borno (CNN 2019). The group's founder was Mohammed Yusuf, a fundamentalist Muslim Cleric who sought to spread Sharia law to all of Nigeria (Ford 2014). Boko Haram's approximate English translation from Hausa, the local language, is "Western education is forbidden" (BBC News 2016). The group's official Arabic name is Jama'atu Ahlis Sunna Lidda'awati wal-Jihad and it translates as "People Committed to the Propagation of the Prophet's Teachings and Jihad" (BBC News 2016).

During the early years of the group, Boko Haram sought to spread its message and gain more followers. In the early days the group was also able to gain funding through donations from sympathetic Nigerians (Brun 2015). During 2003 and 2004, the group participated in sporadic acts of violence against police forces near its headquarters in Maiduguri; in 2005 the group escalated its violence by targeting Christian villages, kidnapping people, and forcibly converting them to Islam (Brun 2015). In response to these attacks, Nigerian police arrested Boko Haram's leader Mohammed Yusuf; Yusuf would later return to the group after making bail (Brun 2015). These limited clashes foreshadowed the violence that would eventually define the group. Kidnapping, in particular, would be one act that would bring the group global notoriety. 2009 would be a pivotal year for the group, and it would create the modern Boko Haram we know today.

2.1.2   The Boko Haram Uprising

In 2009, the violence surrounding Boko Haram escalated to a new level. Government raids and reprisals by Boko Haram would lead to hundreds of deaths. On July 26, 2009, Nigerian Security forces raided one of Boko Haram's safe houses and arrested nine members of the group. In response, Boko Haram attacked several police stations in northeastern Nigeria; an estimated 700 people died in the fighting (Brun 2015). One critical event during this unrest was the capture of Boko Haram's leader Mohammed Yusuf; it is unknown whether he was or was not extrajudicially executed by the police (Brun 2015). The death of Boko Haram's leader would be the catalyst of a decade of violence. After the uprising, Abukarar Shekau became the leader of Boko Haram and he, as well as other key members of the group, fled to nearby Niger, Cameroon, and Chad (Brun 2015). While on the run, the leadership of Boko Haram would regroup and launch one the most destructive modern terrorist campaigns.

2.2 Boko Haram

2.2.1 A Decade of Terror

This section of the chapter highlights some of the heinous acts carried out by Boko Haram in order to paint a picture of the terror campaign over the last 10 years. The events described in this chapter do not represent an exhaustive list of all attacks made by Boko Haram over the last 10 years. After the Boko Haram Uprising, the group was in disarray and armed only with low quality weapons. Boko Haram lacked the resources and technology to wage a sophisticated terror campaign at the end of 2009. However, with help from al-Qaeda, the group began its journey to become one of the most dangerous terrorist groups on the planet.

During late 2010 and early 2011, Boko Haram embarked on a bombing campaign against the government. On New Year's Eve 2010, Boko Haram bombed an army barracks in Abuja, Nigeria, killing 30 people (Associated Press in Abuja 2010). Another one of Boko Haram's devastating early attacks was the bombing of Nigeria's police headquarters. On June 16th, 2011 Boko Haram launched its first suicide bombing when it attacked Nigeria's police headquarters in Abuja (Brock 2011). While the death toll, two, was low, the attack demonstrated the group's appetite for terror. Similar attacks would soon follow. On August 26th, 2011, a suicide bomber detonated a car bomb at the U.N. Headquarters in Abuja: 21 people were killed and 76 were wounded (Nossiter 2011). Throughout the rest of 2011, Boko Haram continued to escalate its terror campaign. During early November 2011, Boko Haram launched a wave of attacks across northern Nigeria. In the bombing and shootings that occurred, over 100 people were killed (Al Jazeera 2011). Similar violence followed throughout 2011 and into 2012.

January 2012 saw several deadly Boko Haram attacks. Throughout the month Boko Haram targeted and killed Christians as well as other civilians. Early in the month, Boko Haram issued an ultimatum to Christians in northern Nigeria telling them to leave the region (Reuters 2012a, b). Boko Haram later followed up on their threats by attacking several towns and killing 37 people; these attacks caused many

Christians to flee the region (Reuters 2012a, b). Later in the month, Boko Haram attacked the city of Kano. The bombing and gunfights left 178 people dead (Reuters 2012a, b). Anti-Christian violence is in line with the group's extremist Islamic stance and would continue throughout 2012.

On April 8th, 2012 Boko Haram launched a bombing attack in Kaduna, Nigeria. The attack killed 38 people and took place near a church (BBC News 2012a, b). The attack came after Boko Haram had threatened violence over Easter (BBC News 2012a, b). Similarly, motivated attacks would follow in June. On June 17th, 2012, multiple churches in the state of Kaduna were bombed by Boko Haram; 36 people died as a result (BBC News 2012a, b). Boko Haram would finish out its 2012 campaign against Christmas with an attack on Christmas Eve. During the holiday, the group shot up and burnt down a church in Peri, a village in the state of Yobe, Nigeria (Al Jazeera 2012). Five people were killed (Al Jazeera 2012). During 2013 Boko Haram would expand beyond attacking Christians by targeting military personnel as well as attacking civilians more indiscriminately.

In March of 2013, Boko Haram carried out a brutal suicide attack on a Bus Depot. On March 18th at least 20 people were killed when a suicide bomber detonated a car bomb (Nossiter 2013a, b). One month later, 200 civilians were massacred in the remote village of Baga; reports detail Nigerian forces burning homes and shooting fleeing civilians (Nossiter 2013a, b). The massacre occurred while security forces were contesting an area controlled by Boko Haram (Nossiter 2013a, b). By this point in time it should be noted that Boko Haram had control of some remote regions in Northeastern Nigeria. As spring became summer, Boko Haram kept up its campaign of terror. On July 6th, 2013, Boko Haram launched an attack on another institution it despises, western style education. In this attack, 42 people were killed after the group shot up, and then burned down, a boarding school in Mamudo, Nigeria (McElroy 2013). Boko Haram actively targets secular education institutions, and fiercely opposes the education of women. Attacks like this are used to intimidate and discourage people from seeking an education.

Violence remained constant throughout the fall of 2013. On September 19th, 159 people were killed in a pair of roadside attacks (Ola 2013). Ten days later, Boko Haram would attack another school with deadly consequences. The group attacked the College of Agriculture in Yobe state and killed students while they were asleep in their dorm; Boko Haram then proceeded to set fire to some of the classrooms (BBC News 2013a, b). As 2013 ended, confrontations between Boko Haram and Nigerian security forces became more regular. One example can be seen in a raid carried out by security forces. On October 25th Nigerian security forces launched a raid on terrorist camps in Borno State. During the operation, the government said they had killed 74 terrorist fighters (Al Jazeera 2013). Four days later, the two groups clashed again. On October 29th security forces battled Boko Haram in a fire fight in Damaturu. During the fighting Boko Haram burned down a police command post as well as an army barracks; in addition, Boko Haram was reported to have looted supplies and vehicles during the raid (Associated Press 2013). After the smoke had cleared 95 terrorists, 23 soldiers, 8 policemen, and 2 civilians had been killed (Associated Press 2013). The violence that occurred during 2013 would fore-shadow the violence of 2014.

To start off 2014, Boko Haram kept up the wholesale killing for which it had become known. At the end of January 2014, the group killed many people in a pair of attacks four days apart. On January 27 Boko Haram bombed, and then torched, the village of Kawuir in Borno state (BBC News 2014a, b, c). In this attack 52 people were reportedly killed (BBC News 2014a, b, c). Four days later, Boko Haram attacked a church in Gulak killing 11 people, adding to death toll of Christians targeted by the group (Godwin 2014). A month later, Boko Haram continued its war on western education. On February 25th 29 boys were murdered at their college in northeastern Nigeria; the group also burned down a dorm at the school (Nossiter 2014). Later in the spring, Boko Haram would commit its most high-profile crime to date.

During the early morning hours of April 15th, 2014, Boko Haram abducted 276 schoolgirls from their school in the town of Chibok (Mbah 2019). Shortly after their abduction, 57 girls were able to escape. This attack brought Boko Haram into the global spotlight as people recoiled in horror at the abduction (Mwai and Goodman 2019). The outrage led to the global social media campaign #BringBackOurGirls with many celebrities and then First Lady Michelle Obama throwing their weight behind it (Mwai and Goodman 2019). Unfortunately, most of the girls who did not escape on the first day would remain in Boko Haram's hands for years, and many are still unaccounted for.

After the Chibok incident, Boko Haram continued its business of slaughter. Through the spring and summer of 2014, the group launched attack after attack. During one attack on May 1st, a car bomb was detonated in the Nyanya suburb of Abuja killing 19 and wounding 60 more (BBC News 2014a, b, c). At the end of the month, Boko Haram attacked a military base amongst other raids. During the attack 24 soldiers and 21 policemen were killed (Associated Press 2014a, b). In June, the group attacked three villages in the Gwoza district of northeastern Nigeria. During these raids it was estimated that more than 200 people had been killed (Associated Press 2014a, b). During this period of 2014, Boko Haram fought consistently with security forces. While Nigerian forces had managed to dislodge Boko Haram from urban areas, the group seized rural areas and villages (Associated Press 2014a, b).

As the fall of 2014 bled into the winter of 2014/2015, Boko Haram did not let up. A bus station bombing in October killed four and left dozens more injured (BBC News 2014a, b, c). Later in December, Boko Haram committed another large-scale kidnapping. On December 14th the group attacked the village Gumsuri; in the attack 32 people were killed and 185 women were abducted (Abubakr and Brumfield 2014). Unfortunately, women abducted in raids like this face horrendous treatment at the hand of Boko Haram. Many women are forcibly married to fighters, while others are raped and sexually assaulted on a daily basis.

As 2015 began, Boko Haram launched one of its deadliest attacks to date. During the week of January 3rd, 2015, Boko Haram overran a Nigerian military base in the town of Baga (BBC News 2015a, b, c). After taking the base, Boko Haram proceeded to torch the town. In the fighting, it was estimated that 2000 people were killed, and 10,000 people displaced; many people traveled across the border into nearby Chad (BBC News 2015a, b, c). At this point in 2015, it was estimated that Boko Haram controlled 70% of the Nigerian state of Borno (BBC News 2015a, b, c). January would mark a horrific first for the group as well. On January, a girl suicide bomber killed herself as well as 19 others in an attack on a Market in Maiduguri (BBC News 2015a, b, c). The girl was estimated to be only ten years old (BBC News 2015a, b, c). Attacks using young female bombers would become commonplace after this attack. Boko Haram often uses young female suicide bombers because they were initially not seen as much as of a threat, and women's clothing makes it easier to conceal a bomb.

Another example of a female suicide bombing can be found less than a month later in February. During a presidential campaign rally, a female suicide bomber blew herself in the city of Gombe (BBC News 2015a, b, c). Violence would remain consistent though the winter and spring of 2015. On March 9th, 2015, another big event occurred. Boko Haram shifted its allegiances away from al-Qaeda and proclaimed that it would be loyal to ISIS (Smith et al. 2015). This move potentially provided the group with more resources and gave Boko Haram renewed international media coverage (Smith et al. 2015). Aligning Boko Haram with ISIS would eventually split the group into two separate factions. After Boko Haram's alliance with ISIS, terror attacks continued in their usual fashion.

March 2015 also saw a change in fortunes for the regional security forces. On March 17th security forces were able to liberate the town of Bama from Boko Haram's control (Akingbule and McGroaty 2015). A few days later it was reported that a multinational force liberated the town of Damasak from the group (Abubakar and Gray 2015). This loss of territory marked the beginning of the end of Boko Haram's control of large swaths of territory. The rest of 2015 would play out similarly to other periods of the insurgency with many bombings, attacks, and, unfortunately, deaths. However, Boko Haram's territorial holdings would be substantially reduced by the end of the year. In early 2016, the government would claim they had defeated Boko Haram, and that no major towns were controlled by the group (AFP 2016).

2016 would start like most other years, with horrible acts of violence, despite claims that Boko Haram was defeated. Just five days after the new year, Boko Haram killed seven during the raiding and bombing of the village of Izgeki (AFP 2016). This killing refutes the government's claims that the group had been truly defeated. A series of events like this would be common in the Boko Haram conflict. Nigerian officials would claim to have defeated the group, only for Boko Haram to retaliate with killings in order to show its potency. Boko Haram kept up the violence with more attacks during the winter of 2016. The group would go on to kill 60 after bombing an Internally Displaced Persons camp in Borno State (Ola 2016). Two female bombers reportedly slipped into the camp before detonating themselves (Ola 2016). The rest of 2016 would see similar violence to past years. Many bombings and raids would claim the lives of hundreds of civilians. However, during 2016 a new trend in Boko Haram's attacks would emerge.

During 2016, Boko Haram would begin to attack and target security forces more frequently. Attacking security forces directly allows Boko Haram to affect the morale of militaries they oppose. Directly engaging with security forces enables the group to refute claims of their defeat, as well as allowing Boko Haram to plunder much needed supplies to sustain its campaign of terror. One example of an attack on security forces can be seen in September. During fighting over the town of Malam Fatori, Boko Haram reportedly killed 40 soldiers after launching a counterattack and forcing the Nigerian military to retreat (Al Jazeera 2016). Another battle with security forces in October would reportedly claim the lives of 20 soldiers (AFP 2016). As 2016 became 2017, Boko Haram's terror campaign would continue its steady march of death and destruction.

During the first week of January 2017, Boko Haram continued to demonstrate its ability to inflict terror in Nigeria. On January 4th three girls, who would have become suicide bombers, were shot and killed by security forces in Maiduguri (Associated Press 2016). Bombings and raids would be common throughout the winter and spring of 2017. On March 30th Boko Haram abducted 22 women in a pair of raids in northeastern Nigeria (France-Presse 2017). Attacks against security forces would continue in 2017 as well. On May 5th, nine Chadian soldiers were killed after Boko Haram attacked the soldiers' positions (AFP 2017a, b). Violence would also affect Cameroon when the group attacked the village of Gakara on August 26th. During the attacks 15 people were killed, eight were abducted, and 30 homes were burned down (Kouagheu 2017).

The closing months of 2017 would see continued violence. On September 13th, a female suicide bomber blew herself up at a mosque in Cameroon; four people were killed (Okogba 2017). Attacks using young female suicide bombers demonstrate Boko Haram's commitment to causing terror at any cost. The early months of 2018 would contain more controversial attacks by the group. On February 19th, 2018, 110 schoolgirls were abducted from their school in Dapchi, Nigeria (Al Jazeera 2017). This attack disputes the Nigerian government's repeated claims that Boko Haram has been technically defeated. Later on, 106 of the girls would be released in March after negotiations with the Nigerian government (BBC News 2018a, b). It was rumored that the government paid Boko Haram a ransom, but the government denies these allegations (BBC News 2018a, b). Another major attack by Boko Haram would occur in May.

On May 1st two suicide bombers attacked the town of Mubi in northeastern Nigeria (Al Jazeera 2018). One bomber blew himself up inside a mosque, while the other detonated his explosives in the market outside. In this attack, 86 people were killed and 58 people were wounded (Al Jazeera 2018). Early in the summer of 2018, another example of Boko Haram's campaign against security forces can be seen. Boko Haram attacked soldiers posted in the town of Gajiram; 9 were killed (The New Indian Express 2018). Violence continued throughout the summer and early fall of 2018. However, in November the Nigerian military would suffer an embarrassing defeat. A military base in Metele, Nigeria, was overrun by Boko Haram (Haruna 2018). Seventy soldiers were killed in the assault and soldiers claim the defeat occurred because they were underequipped (Haruna 2018). Boko Haram was also able to make off with several gun trucks in the raid (Haruna 2018). Violence would continue throughout the remainder of 2018.

As 2019 began the Boko Haram insurgency entered its tenth year. As of the writing of this book, Boko Haram is still actively waging its campaign of terror in Northeastern Nigeria, as well as the region around Lake Chad. Despite repeated claims that the group has been defeated by Nigerian forces, the group is able to inflict casualties on civilians and the military alike. The road to defeating Boko Haram is long and winding but we hope the information presented in later chapters can help the Government of Nigeria and its neighbors as well as relevant international stakeholders tip the balance.

2.2.2 Timeline of Events

This timeline contains all of the events discussed in the previous sections. It is by no means an exhaustive list of all of the events involving Boko Haram over the last 20 years.

- **2002:** Boko Haram is founded in Maiduguri, Nigeria (Ford 2014).
- **July 26, 2009:** The Boko Haram uprising occurs, and Mohammed Yusuf is killed. Seven-hundred are killed in the violence (Brun 2015).
- **December 31, 2010:** Thirty people are killed when Boko Haram bombs an army barracks in Abuja (Associated Press in Abuja 2010).
- **June 16, 2011:** Boko Haram bombs Nigeria's police headquarters in Abuja (Brock 2011).
- **August 26, 2011:** The Nigerian U.N. headquarters was bombed killing 21 people (Nossiter 2011).
- **November 4, 2011:** Boko Haram launches a wave of coordinated attacks killing 100 (Al Jazeera 2011).
- **January 2012:** Boko Haram launches several attacks killing many Christians as well as other civilians (Reuters 2012a, b).
- **April 8, 2012:** Boko Haram bombs the city of Kaduna during the Easter holiday killing 38 (BBC News 2012a, b).
- **June 17, 2012:** Several churches are bombed by Boko Haram killing 36 (BBC News 2012a, b).
- **December 24, 2012:** Boko Haram attacks a church killing five (Al Jazeera 2012).
- **March 18, 2013:** Boko Haram bombs a bus depot killing over 20 (Nossiter 2013a, b).
- **April 16, 2013:** Nigerian security forces massacre an estimated 200 people in the village of Baga (Nossiter 2013a, b).
- **July 6, 2013:** After an attack on a boarding school 42 people are dead (McElroy 2013).

- **September 19, 2013:** 159 people are killed in two attacks by Boko Haram (Ola 2013).
- **September 29, 2013:** Boko Haram attacks the College of Agriculture in Yobe state killing 50. (BBC News 2013a, b).
- **October 25, 2013:** Nigerian forces raid terrorist camps killing 74 enemy combatants (Al Jazeera 2013).
- **October 29, 2013:** 129 people are killed after a terrorist raid on Damaturu (Associated Press 2013).
- **January 27, 2014:** Boko Haram bombs and sets fire to the village of Kawuri killing 52 (BBC News 2014a, b, c).
- **January 31, 2014:** A church in Gulak was attacked by Boko Haram and 11 people were killed (Godwin 2014).
- **February 25, 2014:** Twenty students were killed after an attack on a college in northeastern Nigeria (Nossiter 2014).
- **April 15, 2014:** 276 girls were abducted from their school in Chibok Nigeria. 57 girls managed to escape almost immediately leaving 219 in the hands of the group. This attack led to the creation of the worldwide social media campaign #BringBackOurGirls (Mbah 2019).
- **May 1, 2014:** A car bombing occurs in a suburb of Abuja killing 19 people (BBC News 2014a, b, c).
- **May 27, 2014:** After an attack on a military installation 24 soldiers and 21 policemen are dead (Associated Press 2014a, b).
- **June 2, 2014:** Over 200 people are killed by Boko Haram during attacks in the Gwoza government district (Associated Press 2014a, b).
- **October 31, 2014:** A bus station bombing in Gombe, Nigeria, leaves 4 dead and 32 wounded (BBC News 2014a, b, c).
- **December 13, 2014:** Thirty-two people are killed and 185 are abducted after Boko Haram attacks the village of Gumsuri (Abubakr and Brumfield 2014).
- **January 3, 2015:** Boko Haram overruns a military base in the town of Baga. Afterwards the group razed the town killing an estimated 2000 people and displacing 10,000 (BBC News 2015a, b, c).
- **January 10, 2015:** Boko Haram forces a young girl to be a suicide bomber and 19 are killed in the attack (BBC News 2015a, b, c).
- **February 2, 2015:** A female suicide bomber kills one during an attack on a Nigerian presidential campaign rally (BBC News 2015a, b, c).
- **March 9, 2015:** Boko Haram proclaims its allegiance to ISIS (Smith et al. 2015).
- **March 17, 2015:** The Nigerian military reclaims the town of Bama from the hands of Boko Haram (Akingbule and McGroaty 2015).
- **March 21, 2015:** Damasak is liberated from Boko Haram's control. These liberations mark the beginning of the end of Boko Haram control of large areas of land (Abubakar and Gray 2015).
- **January 2016:** The Nigerian Government claims Boko Haram has been defeated after retaking most of the land seized by the group (AFP 2016).

- **January 6, 2016:** Boko Haram attacks the village of Izgeki killing 7. This attack shows that despite the Nigerian government's claims, the group has the capacity to kill (AFP 2016).
- **February 10, 2016:** Two female suicide bombers attack an Internally Displaced Persons camp in Borno State killing 60 people (Ola 2016).
- **September 22, 2016:** An estimated 40 soldiers are killed in an engagement with Boko Haram (Al Jazeera 2016).
- **October 17, 2016:** Boko Haram claims responsiblity for the 20 soldiers killed in a firefight (AFP 2016).
- **January 4, 2017:** Three girl suicide bombers are killed in Nigeria (Abdulaziz 2017).
- **March 30, 2017:** Boko Haram abducted 22 women in northeastern Nigeria (France-Presse 2017).
- **May 5, 2017:** Nine soldiers are killed after Boko Haram attacks a Chadian military base in the Lake Chad region (AFP 2017a, b).
- **August 26, 2017:** Boko Haram attacked the village of Gakara. Fifteen people were killed, 8 were abducted, and many homes were torched (Kouagheu 2017).
- **September 13, 2017:** A female Boko Haram suicide bomber kills 4 in Cameroon (Okogba 2017).
- **February 19, 2018:** 110 girls are kidnapped from their school in Dapchi, Nigeria (Al Jazeera 2017).
- **March 21, 2018:** Boko Haram releases 106 of the girls abducted from Dapchi after negotiating with the Nigerian government (BBC News 2018a, b).
- **May 1, 2018:** A dual suicide bombing kills 86 in Mubi, Nigeria (Al Jazeera 2018).
- **June 20, 2018:** Boko Haram kills 9 soldiers in an attack on their positions in Gajiram (The New Indian Express 2018).
- **November 22, 2018:** Boko Haram overruns a military base in Metele, Nigeria, killing 70 soldiers (Haruna 2018).

2.3 Organizational Overview

2.3.1 Structure and Finance

Currently there are two separate factions of Boko Haram operating in northeastern Nigeria. One of these factions is allied to ISIS, while the other is led by Abubakar Shekau, the group's leader since 2009 (BBC News 2018a, b). The split occurred after the group pledged allegiance to ISIS in March 2015. Tensions developed between Shekau and ISIS leadership, which caused ISIS to name a new leader for Boko Haram. Shekau, unwilling to accept this leadership change broke away and formed his own sect separate from the portion of the group aligned with ISIS (Counter Extremism Project 2019).

While there are currently two separate factions of Boko Haram, the group operates with a decentralized leadership structure. Before the splitting of the group, Shekau was the de facto leader. Below the leader are the 30 members of the Shura council; members of the council have command over the many cells that make up Boko Haram (Stratfor 2014). Individual cells within Boko Haram are often specialized. Some cells consist of combat troops, kidnapping specialists, or explosive experts. Other cells provide critical support functions like intelligence gathering, the procurement of supplies, medical administration, and recruitment (Stratfor 2014). One of the benefits of this operational structure is information security. Individual cells and members of the Shura council usually do not know what other cells' operational goals are, and this prevents the leakage of information if one cell were to be compromised (Stratfor).

Boko Haram reportedly receives some of its funding from other Islamic terrorist groups. For example, in its early days the group was reported to have close ties with al-Queda. Boko Haram even went so far as to say that "al Qaeda are our elder brothers" (McCoy 2014). This relationship between the two groups supports the reports that up to $3 million was funneled to Islamic fundamentalist groups in Northern Nigeria in the early 2000s (McCoy 2014). The group also reportedly receives funding from al-Shabaab in Somalia.

Another large portion of Boko Haram's funding comes from kidnappings and other human trafficking activities. In April of 2013 a French family was freed after a ransom of $3 million was reportedly paid to the group (BBC News 2013a, b). The group also threatened to sell the schoolgirls kidnapped from Chibok on the black market (McCoy 2014). The way Boko Haram operates allows this illicit money to be used to its maximum potential.

Boko Haram's low-tech operations allow it to get the most out of every illicit dollar. AK-47 s, 7.62 x 39 mm ammunition, rocket propelled grenades are all cheap commodities (Stewart and Wroughton 2014). Furthermore, the group often resupplies itself through looting villages and military outposts. Most of the group's military supplies have been stolen from the Nigerian government. For example, in November of 2018 Boko Haram attacked a military base in Metele, a village in Northeastern Nigeria, and made off with four tanks as well as other support vehicles (Reinl 2019). During attacks on villages Boko Haram often loots food supplies, livestock, and robs banks. It is estimated that bank robberies account for $6 million of the group's assets (McCoy 2014). Raids like these allow the group to resupply and maintain its operational capacity.

2.3.2 Areas of Operation

Boko Haram's operational area currently includes four countries in western Africa, Nigeria, Niger, Chad and Cameroon. For most of the first five years of its campaign of terror, Boko Haram launched attacks in central and northeastern Nigeria. Starting in 2014, the group expanded its area of operations into some of Nigeria's neighbors.

Southern Niger, southeastern Chad, and northern Cameroon all began to suffer from Boko Haram's campaign of terror. As of 2019, the group has been very active in the Lake Chad Basin and has been launching regular attacks into Nigeria, Chad, Cameroon, Niger.

2.3.3 Membership

The membership of Boko Haram is comprised mostly of Nigerian men from the northeastern region of the country. Boko Haram is able to exploit the geographic features of the region in order to recruit new members. Nigerian population is split along both religious and economic lines. Boko Haram is able to recruit its membership from the predominantly Muslim, and poverty stricken, north of the country (Congressional Research Service 2018). Young men are very susceptible to Boko Haram's call to arms because of their lack of economic opportunity. It is estimated that youth in the 15–24 age group have an unemployment rate of 12.4% and 30.8% of men are illiterate (World Factbook 2019). Young men with no better options often see Boko Haram's financial promises as the best offer they will get (Counter Extremism Project 2019).

Boko Haram's members also consist of those forced or abducted into its ranks. Over the years it has been estimated that 10,000 boys have been kidnapped by Boko Haram; these boys are then trained to become foot soldiers for the group (Hinshaw and Parkinson 2016). Boys as young as five learn the ins and outs of shooting rifles, while others are trained to be suicide bombers (Hinshaw and Parkinson 2016). Boko Haram abducts these members by seizing a village and shooting any boy or teenager who does not join the group (Counter Extremism Project 2019). Young girls are not exempt from Boko Haram kidnappings. They are taken by the group to be used a sex slaves or forced to become suicide bombers; in early 2018 it was estimated that 1200 people had been killed by female suicide bombers during the conflict (Zenn 2018a, b).

Overall, the members of Boko Haram consist mostly of those who are forced to fight under the penalty of death or see no better options for themselves. Those who seek to leave the group face dire consequences if discovered. In 2018, one of the group's leaders, Ali Gaga, was executed for suspected disloyalty; Gaga was killed for being suspected of wanting to surrender to Nigerian military forces (The Commonwealth 2019). This example shows how fear and intimidation can effectively be used by the group to keep people in line. Furthermore, for the people who do manage to leave the group, social stigma prevents them from living normal lives.

Women and girls who escape the group, surrender while on suicide missions, or are rescued by security forces, are often rejected by their families and communities. They are seen as liabilities because of their relationship with Boko Haram (Bradford 2017). Women who become pregnant due to sexual violence face a similar stigma. The children of these women are also ostracized because communities consider them to carry the "'bad blood'" of Boko Haram (Bradford 2017). Women and children are not the only ones affected by this social isolation.

The young boys forced to fight for Boko Haram face similar treatment to the women who escape from the group. Like their female counterparts, boys taken by the group are seen as dangerous and undesirable (Dwyer 2017). One boy who escaped from Boko Haram reportedly lives on the streets of Maiduguri, the capital of Borno State in Nigeria. He lives in fear that discovery of his past with Boko Haram will make his life even harder (Dwyer 2017). Other boys who have escaped fear being discovered by the military even though amnesty programs exist (Hinshaw and Parkinson 2016). The stigma surrounding survivors of Boko Haram is a long-term obstacle in fixing some of the damage caused by the group.

2.3.4 Tactics

Over the years, Boko Haram's tactics have evolved with the ever-changing conditions in northeastern Nigeria. Abductions, suicide bombings, the targeting of military personnel, and looting form the key components of Boko Haram's playbook of terror. Currently, the sect of Boko Haram that is allied to ISIS has been targeting and attacking military targets in the rural regions of northeastern Nigeria (Thurston 2017). Raiding military installations allows the group to accomplish multiple things at once. Attacks on military bases not only degrade Nigerian operational capacity but also allow Boko Haram to replenish their own supplies. For example, raids in the autumn of 2017 saw Boko Haram seize caches of weapons, ammunition, as well as anti-aircraft weapons mounted on trucks (Thurston 2017). Raids on military bases are also designed to erode the morale of those serving in the military. The group targets villages in its raids for similar reasons. By looting villages, the group is able to seize critical food and supplies (Thurston 2017). Overall, the group's campaign of hit-and-run raids has caused significant casualties and loss of material for the Nigerian military.

Boko Haram has also consistently used suicide bombings in its war of terror. The first Boko Haram suicide bombing occurred in 2011. A recent bombing during June of 2019 caused the deaths of 30 people (France-Presse 2019). The bombing took place in a local hall near the Nigerian regional capital of Maiduguri. Attacks like this have become a hallmark of the group. The group uses suicide bombing to cause terror throughout northeastern Nigeria. Boko Haram is also known for its ruthless use of women and children as bombers. The group started using female suicide bombers in 2014, before this year all of the bombers had been men (Meservey and Vadyak 2018). Female suicide bombers are used for multiple reasons. The group views women as expendable in comparison to the group's male combatants; in addition, women are perceived to be less of a threat, while their clothing is better able to conceal bombs (Meservey and Vadyak 2018). Most of these female suicide bombers were abducted by the group, and some see becoming a suicide bomber as a viable way to escape the grotesque treatment they receive at the hands of the group (The Mainichi 2018).

Women are not the only group forced to become suicide bombers. Boko Haram is also notorious for its use of child-bombers. Similar to women, children are used as suicide bombers because they raise fewer red flags with security forces, and the group views them as expendable. For example, from 2011 to 2017, 134 suicide bomber's ages were able to be determined: 80 suicide bombers out of this group were either teenagers or children (Kriel 2017). The youngest of these bombers was estimated to be seven years old (Kriel 2017). Boko Haram's wholesale use of women and children is unprecedented, and these deplorable tactics demonstrate how far the group is willing to go in its pursuit to establish Islamic rule in Nigeria.

Abductions are another popular tool of Boko Haram. By abducting people, Boko Haram is able destroy communities as well as instill fear in the populace. However, abductions also provide the group with people who can be forced to fight for the group. Children can be trained to be fighters or suicide bombers. Women are often forcibly wed to fighters or trained to be suicide bombers. Abductees of significant value can be held for ransom to finance the group's operations. All of these outcomes strengthen the group and allow it to continue its terrorist operations. One example can be seen in the Chibok kidnaping. In April of 2014 over 100 girls were taken from their school in Chibok Nigeria; over three years later the Nigerian Government released five Boko Haram commanders and reportedly gave the group €2 million in cash (Parkinson and Hinshaw 2017). Abductions continue to be used effectively by the group and make life perilous in the remote portions of northeastern Nigeria.

2.3.5 Training

Boko Haram's training practices are relatively opaque. The group has not revealed much information on how it trains the members it recruits or abducts. It is assumed that the group has training sites outside of Nigeria in Cameroon and Somalia (Counter Extremism Project 2019). It is also expected that the group has received training from advisors from other terrorist groups. Internally, it is expected that members of the groups are taught the basics of handling small arms. It is also expected that some members are trained to build bombs in order for the group's suicide bombing campaign to continue. However, some anecdotes reveal how some of the training of child soldiers occurs.

Boys who managed to escape the group recall their time with Boko Haram. The boys were taught how to shoot and operate AK-47 s. Individual boys recall being taught to shoot at sandbags as well as prisoners sentenced to death by the group (Topol 2017). The group also uses violence and social engineering to change the behavior of boys. By forcing the boys to be involved acts of terrible violence, the children's ability to relate to their old life at home diminishes and they become desensitized to violence (Topol 2017). By normalizing violence, Boko Haram is able to make their child soldiers into more efficient killers.

2.4   Relationships with Other Terrorist Groups

2.4.1   al-Qaeda

Boko Haram's relationship with al-Qaeda began during the early days of the group. Reportedly, Boko Haram's founder Muhammed Yusuf had met Osama bin Laden in the 1990s (Zenn 2018a, b). This relationship allowed Boko Haram to gain valuable training and operational skills from the more seasoned terrorist organization. By working with al-Qaeda, Boko Haram was able to gain weapons and bomb making training in addition to the capital needed to fund its terrorist activities (Staffell and Awan 2016). It is also hypothesized that al-Qaeda provided direction in Boko Haram's early suicide bombing attacks; the attacks mirrored those al-Qaeda in the early 2010s (Zenn 2018a, b). Under the leadership of Abubakar Shekau, Boko Haram's relationship with al-Qaeda began to deteriorate in 2011 and in 2014 al-Qaeda cut ties with the Nigerian terrorist group (Zenn 2018a, b).

2.4.2   ISIS

Boko Haram's formal relationship with ISIS began in 2015. During March of this year, Boko Haram's leader Abubakar Shekau aligned his group with ISIS (Staffell and Awan 2016). This marked a departure from the group's previous relationship with al-Qaeda. Boko Haram's alliance with ISIS gave the group renewed support and access to the materials and money need to wage its war of terror. However, this relationship with ISIS would lead to a lasting split in the group. Conflicts between the leadership of ISIS and Shekau lead to ISIS appointing a new leader for Boko Haram in 2016 (BBC News 2018a, b). Shekau's faction operates separately from the ISIS backed faction of Boko Haram. The ISIS affiliated faction is also known as the Islamic State's West African province or ISWA (BBC News 2018a, b). The current leader of the ISWA faction is believed to be Ibn Umar al-Barnawi (Kelly 2019). During the March of 2019, the last Middle Eastern village controlled by the Islamic State was liberated and the group's caliphate was declared dead (Callimachi 2019). The fall of ISIS leaves the group's relationship up in the air. However, the two major factions of Boko Haram remain.

References

Abdulaziz I (2017) Three girl suicide bombers are killed in Nigeria, Washington Times, Jan 4 2017. https://www.washingtontimes.com/news/2017/jan/4/3-girl-suicide-bombers-gunned-down-in-northeast-ni/

Abubakar A, Gray M (2015) Mass grave found in former Boko Haram-held town in Nigeria. CNN. https://edition.cnn.com/2015/03/20/africa/nigeria-mass-grave/. Accessed 28 July 2019

Abubakr A, Brumfield B (2014) Officials: Boko Haram kidnaps 185 women and children, kills 32 people. CNN. https://edition.cnn.com/2014/12/18/world/africa/nigeria-boko-haram-kidnapping/. Accessed 23 July 2014

AFP (2016) Boko Haram kills seven in suicide attack, raid: residents. The Guardian. https://guardian.ng/news/boko-haram-kill-seven-in-suicide-attack-raid-residents/. Accessed 29 July 2019

AFP (2017a) Boko Haram claims attack on soldiers in NE Nigeria. Daily Mail.com. https://www.dailymail.co.uk/wires/afp/article-3845112/Boko-Haram-claims-attack-soldiers-NE-Nigeria.html. Accessed 31 July 2019

AFP (2017b) Boko Haram attack in Chad kills nine troops, 40 jihadists: sources. Yahoo!. https://www.yahoo.com/news/boko-haram-attack-chad-kills-nine-troops-40-182816218.html. Accessed 31 July 2019

Akingbule G, McGroaty P (2015) Boko Haram driven from strategically important Bama, says Nigerian Army. The Wall Street Journal. https://www.wsj.com/articles/boko-haram-driven-from-strategically-important-bama-says-nigerian-army-1426605335. Accessed 28 July 2019

Al Jazeera (2011) Nigeria group threatens more deadly attacks. https://www.aljazeera.com/news/africa/2011/11/20111169858380467.html. Accessed 21 July 2019

Al Jazeera (2012) Deadly Christmas attack on Nigeria church. https://www.aljazeera.com/news/africa/2012/12/2012122514557537669.html. Accessed 21 July 2019

Al Jazeera (2013) Scores killed in Nigeria military raid. https://www.aljazeera.com/news/africa/2013/10/scores-killed-nigeria-military-raid-2013102517230225814.html. Accessed 21 July 2019

Al Jazeera (2016) Nigeria: Troops battle Boko Haram near Malam Fatori. https://www.aljazeera.com/news/2016/09/nigeria-troops-battle-boko-haram-malam-fatori-160921153240745.html. Accessed 31 July 2019

Al Jazeera (2017) 110 Nigerian schoolgirls still missing after attack: Minister. https://www.aljazeera.com/news/2018/02/110-nigerian-schoolgirls-missing-attack-minister-180225171154082.html. Accessed 4 Aug 2019

Al Jazeera (2018) Nigeria mosque attack death toll rises to 86. https://www.aljazeera.com/news/2018/05/nigeria-mosque-attack-death-toll-rises-86-180502162539760.html. Accessed 4 Aug 2019

Associated Press (2013) 128 dead in Islamic extremist attack on Nigeria state capital in state of emergency. Fox News. https://www.foxnews.com/world/128-dead-in-islamic-extremist-attack-on-nigeria-state-capital-in-state-of-emergency. Accessed 21 July 2019

Associated Press (2014a) Boko Haram Militants Dressed as Soldiers Slaughter Scores: Witnesses. NBC News. https://www.nbcnews.com/news/world/boko-haram-militants-dressed-soldiers-slaughter-scores-witnesses-n123251. Accessed 23 July 2019

Associated Press (2014b) Nigerian security forces hit in fresh Boko Haram attack. The Telegraph. https://www.telegraph.co.uk/news/worldnews/africaandindianocean/nigeria/10859197/Nigerian-security-services-hit-in-fresh-Boko-Haram-attack.html. Accessed 23 July 2019

Associated Press (2016) 3 girl suicide bombers gunned down in Nigeria. CBC. https://www.cbc.ca/news/world/boko-haram-suicide-bombing-1.3921989. Accessed 31 July 2019

Associated Press in Abuja (2010) Nigerian bomb blast hit army barracks during New Year celebrations. The Guardian. https://www.theguardian.com/world/2010/dec/31/nigeria-bomb-blast-abuja. Accessed 21 July 2019

BBC News (2012a) Nigeria: Dozens dead in church bombings and rioting. https://www.bbc.com/news/world-africa-18475853. Accessed 21 July 2019

BBC News (2012b) Nigerian Easter bomb kills many in Kaduna. https://www.bbc.com/news/world-17650542. Accessed 212 July 2019

BBC News (2013a) Nigeria attack: students shot dead as they slept. https://www.bbc.com/news/world-africa-24322683. Accessed 21 July 2019

BBC News (2013b) Nigeria's Boko Haram 'got $3m ransom' to free hostages. https://www.bbc.com/news/world-africa-22320077. Accessed 16 June 2019

BBC News (2014a) Abuja blast: car bomb attack rocks Nigerian capital. https://www.bbc.com/news/world-africa-27249097. Accessed 24 July 2014

BBC News (2014b) Nigeria 'Boko Haram' attacks leave scores dead. https://www.bbc.com/news/world-africa-25916810. Accessed 23 July 2019

BBC News (2014c) Nigeria bus station blast during rush hour. https://www.bbc.com/news/world-africa-29849629. Accessed 28 July 2019

BBC News (2015a) Boko Haram crisis: Nigeria's Baga town hit by new assault. https://www.bbc.com/news/world-africa-30728158. Accessed 28 July 2019

BBC News (2015b) Nigerian elections: blast hits presidential rally in Gombe. https://www.bbc.com/news/world-africa-31099348. Accessed 28 July 2019

BBC News (2015c) Nigeria: 'Girl bomber' kills 19 people in Maiduguri market. https://www.bbc.com/news/world-africa-30761963. Accessed 28 July 2019

BBC News (2016) Who are Nigeria's Boko Haram Islamist group? https://www.bbc.com/news/world-africa-13809501. Accessed 10 July 2019

BBC News (2018a) Boko Haram in Nigeria: split emerges over leadership. https://www.bbc.com/news/world-africa-36973354. Accessed 16 June 2019

BBC News (2018b) Dapchi girls: freed Nigerian girls tell of kidnap ordeal. https://www.bbc.com/news/43489217. Accessed 4 Aug 2019

Bradford A (2017) Former Boko Haram captives face Stigma, often from other survivors. Huffington Post. https://www.huffpost.com/entry/boko-haram-survivors-face-stigma_n_58e7f434e4b058f0a02f3c0f?guccounter=1&guce_referrer=aHR0cHM6Ly93d3cuZ29vZ2xlLmNvbnNB8&guce_referrer_sig=AQAAAMTqRjR5dEEMxoHcP8mapsBIwAUqrrbvng7GmXfKqg-WIRz82pLCJQ3jAGuxyqYSfGYJw0ggKixgnwnU3-9vcN93mkuOK6Rr5Rjo1cWzzgaptzC028wEmfoleKuXjJoxE_1ckNK6B1ZX1LLVEf0ik6OUFFJYRtv21_S4iW2SU9C2a. Accessed 22 June 2019

Brock J (2011) Nigerian Islamist sect claims bomb attack: paper. Reuters. https://af.reuters.com/article/topNews/idAFJOE75G0BF20110617?sp=true. Accessed 21 July 2019

Brun H (2015) Boko Haram: the early years. The Interpreter. https://www.lowyinstitute.org/the-interpreter/boko-haram-early-years. Accessed 11 July 2019

Callimachi R (2019) ISIS caliphate crumbles as last village in Syria Falls. The New York Times. https://www.nytimes.com/2019/03/23/world/middleeast/isis-syria-caliphate.html. Accessed 8 July 2019

Campbell J (2011) Why Nigeria's North South distinction is important. The Huffington Post. https://www.huffpost.com/entry/why-nigerias-north-south_b_817734. Accessed 10 July 2019

CNN (2019) Boko Haram fast facts. https://www.cnn.com/2014/06/09/world/boko-haram-fast-facts/index.html. Accessed 10 July 2019

Congressional Research Service (2018) Boko Haram and the Islamic State's West Africa Province. https://fas.org/sgp/crs/row/IF10173.pdf. Accessed 16 June 2019

Counter Extremism Project (2019) Boko Haram. https://www.counterextremism.com/threat/boko-haram. Accesses 16 June 2019

Dwyer H (2017) In Nigeria, one boy's survival at the hands of Boko Haram and his long journey home. Unicef. https://www.unicef.org/nigeria/stories/nigeria-one-boys-survival-hands-boko-haram-and-his-long-journey-home. Accessed 22 July 2019

Ford J (2014) The Origins of Boko Haram. The National Interest. https://nationalinterest.org/feature/the-origins-boko-haram-10609. Accessed 10 July 2019

France-Presse A (2017) Boko Haram kidnaps 22 girls and women in North-East Nigeria. The Guardian. https://www.theguardian.com/world/2017/apr/01/boko-haram-kidnaps-22-girls-and-women-in-north-east-nigeria. Accessed 31 July 2019

France-Presse A (2019) Dozens killed in triple suicide Bombing in Nigeria. The New York Times. https://www.nytimes.com/2019/06/17/world/africa/nigeria-attack.html. Accessed 24 June 2019

Godwin A (2014) Gunmen invade church, kill pastor, 10 worshippers in Adamawa. Daily Post. https://dailypost.ng/2014/02/02/gunmen-invade-church-kill-pastor-10-worshippers-adamawa/. Accessed 23 July 2019

Haruna A (2018) How Boko Haram killed 'over 70 soldiers' in Metele attack -survivor. The Premium Times. https://www.premiumtimesng.com/regional/nnorth-east/296967-how-boko-haram-killed-over-70-soldiers-in-metele-attack-survivor.html. Accessed 4 Aug 2019

Hinshaw D, Parkinson J (2016) The 10,000 kidnapped boys of Boko Haram. The Wall Street Journal. https://www.wsj.com/articles/the-kidnapped-boys-of-boko-haram-1471013062. Accessed 18 June 2019

Kelly F (2019) Boko Haram of Islamic State West Africa … or both? The Defense Post. https://thedefensepost.com/2019/02/01/boko-haram-islamic-state-west-africa/. Accessed 8 July 2019

Kouagheu J (2017) Suspected Boko Haram militants kill 15 in Cameroon. Reuters. https://af.reuters.com/article/topNews/idAFKCN1B6078-OZATP. Accessed 31 July 2019

Kriel R (2017) Boko Haram favors women, children as suicide bombers, study reveals. CNN. https://www.cnn.com/2017/08/10/africa/boko-haram-women-children-suicide-bombers/index.html. Accessed 29 June 2019

Mbah F (2019) Nigeria's Chibok schoolgirls: five years on, 112 still missing. Al Jazeera. https://www.aljazeera.com/news/2019/04/nigeria-chibok-school-girls-years-112-missing-190413192517739.html. Accessed 23 July 2019

McCoy T (2014) This is how Boko Haram funds its evil. The Washington Post. https://www.washingtonpost.com/news/morning-mix/wp/2014/06/06/this-is-how-boko-haram-funds-its-evil/?utm_term=.da866cc82ba5. Accessed 16 June 2019

McElroy D (2013) Extremist attack in Nigeria kills 42 at boarding school. The Telegraph. https://www.telegraph.co.uk/news/worldnews/africaandindianocean/nigeria/10163942/Extremist-attack-in-Nigeria-kills-42-at-boarding-school.html. Accessed 21 July 2019

Meservey J, Vadyak A (2018) Boko Haram's Sick ploy to turn girls into suicide bombers. The Heritage Foundation. https://www.heritage.org/terrorism/commentary/boko-harams-sick-ploy-turn-girls-suicide-bombers. Accessed 27 June 2019

Mwai P, Goodman J (2019) Nigerian elections: is poverty getting worse? BBC News. https://www.bbc.com/news/world-africa-47122411. Accessed 10 July 2019

Nossiter A (2011) Islamic group says it was behind fatal Nigeria attack. The New York Times. https://www.nytimes.com/2011/08/29/world/africa/29nigeria.html Accessed 21 July 2019

Nossiter A (2013a) Bombs strike bus station in Nigeria. The New York Times. https://www.nytimes.com/2013/03/19/world/africa/suicide-bombers-strike-bus-depot-in-nigeria.html. Accessed 21 July 2019

Nossiter A (2013b) Massacre in Nigeria Spurs Outcry Over Military Tactics. The New York Times. https://www.nytimes.com/2013/04/30/world/africa/outcry-over-military-tactics-after-massacre-in-nigeria.html. Accessed 21 July 2019

Nossiter A (2014) Islamist militant blamed for deadly college attack in Nigeria. The New York Times. https://www.nytimes.com/2014/02/26/world/africa/dozens-killed-in-nigeria-school-assault-attributed-to-islamist-militant-group.html?_r=0. Accessed 23 July 2019

Okogba E (2017) Four killed in suicide attack in Cameroon's restive North. Vanguard. https://www.vanguardngr.com/2017/09/four-killed-suicide-attack-cameroons-restive-north/. Accessed 4 Aug 2019

Ola L (2013) Nigerian Islamists kill at least 159 in two attacks. Reuters. https://www.reuters.com/article/us-nigeria-violence-toll/nigerian-islamists-kill-at-least-159-in-two-attacks-idUSBRE98J0SP20130920. Accessed 21 July 2019

Ola L (2016) Female suicide bombers kill over 60 in northeast Nigeria: officials. Reuters. https://www.reuters.com/article/us-nigeria-violence-idUSKCN0VJ265. Accessed 29 July 2019

Oyekunle O (2019) The largest economies in Africa by GDP, 2019. The African Exponent. https://www.africanexponent.com/post/9786-top-six-countries-with-the-biggest-gdp-in-africa. Accessed 9 July 2019

Parkinson J, Hinshaw D (2017) Freedom for the World's most famous hostages came at heavy price. The Wall Street Journal. https://www.wsj.com/articles/two-bags-of-cash-for-boko-haram-the-untold-story-of-how-nigeria-freed-its-kidnapped-girls-1513957354. Accessed 29 June 2019

Reinl J (2019) How stolen weapons keep groups like Boko Haram in business. PRI. https://www.pri.org/stories/2019-04-19/how-stolen-weapons-keep-groups-boko-haram-business. Accessed 16 June 2019

Reuters (2012a) Christians flee attacks in Northeast Nigeria. https://web.archive.org/web/20120113224320/http://www.trust.org/alertnet/news/christians-flee-attacks-in-northeast-nigeria/. Accessed 21 July 2019

Reuters (2012b) Islamist insurgents kill over 178 in Nigeria' Kano. https://archive.is/20130416020926/http://www.trust.org/alertnet/news/islamist-insurgents-kill-over-178-in-nigerias-kano/. Accessed 21 July 2019

Smith M, Smyth R, Windrem R (2015) What does Boko Haram's pledge of allegiance to ISIS really mean. NBC News. https://www.nbcnews.com/storyline/isis-terror/what-does-boko-harams-isis-allegiance-pledge-really-mean-n319856. Accessed 28 July 2019

Staffell S. and Awan A. (2016) Jihadism transformed: Al-Qaeda and Islamic State's global battle of ideas. Oxford Scholarship Online. https://www.oxfordscholarship.com/view/10.1093/acprof:oso/9780190650292.001.0001/acprof-9780190650292-chapter-008. Accessed 8 July 2019

Stewart P, Wroughton L (2014) How Boko Haram is beating U.S. efforts to choke its financing. Reuters. https://www.reuters.com/article/us-usa-nigeria-bokoharam-insight/how-boko-haram-is-beating-u-s-efforts-to-choke-its-financing-idUSKBN0F636920140701. Accessed 16 June 2019

Stratfor (2014) Nigeria: examining Boko Haram. https://worldview.stratfor.com/article/nigeria-examining-boko-haram. Accessed 16 June 2019

Tertsakian C (2004) "Ploitical Shari'a?" Human Rights and Islamic Law in Northern Nigeria. Human Rights Watch. https://www.hrw.org/report/2004/09/21/political-sharia/human-rights-and-islamic-law-northern-nigeria. Accessed 9 July 2019

The Commonwealth (2019) Nigeria: history. http://thecommonwealth.org/our-member-countries/nigeria/history. Accessed 10 June 2019

The Mainichi (2018) Marriage or slavery? For girls abducted by Boko Haram, suicide bombing an escape. https://mainichi.jp/english/articles/20180404/p2a/00m/0na/006000c. Accessed 29 June 2019

The New Indian Express (2018) Boko Haram kills nine soldiers in Nigeria. http://www.newindianexpress.com/world/2018/jun/20/boko-haram-kills-nine-soldiers-in-nigeria-1830875.html. Accessed 4 Aug 2019

Thurston A (2017) Boko Haram's New Tactics Imperil Nigeria's Countryside. IPI Global Observatory. https://theglobalobservatory.org/2017/11/boko-harams-new-tactics-imperil-countryside/. Accessed 22 June 2019

Topol S (2017) Trained to kill: how four boy soldiers survived Boko Haram. The New York Times. https://www.nytimes.com/2017/06/21/magazine/boko-haram-the-boys-from-baga.html. Accessed 29 June 2019

World Factbook (2019) Nigeria. Central Intelligence Agency. https://www.cia.gov/library/publications/the-world-factbook/geos/ni.html Accessed 16 June 2019

Zenn J (2018a) Boko Haram's al-Qaeda affiliation: a response to 'Five Myths about Boko Haram'. Lawfare. https://www.lawfareblog.com/boko-harams-al-qaeda-affiliation-response-five-myths-about-boko-haram. Accessed 8 July 2019

Zenn J (2018b) Boko Haram beyond the headlines: analyses of Africa's enduring insurgency. Combating Terrorism Center at West Point. https://ctc.usma.edu/app/uploads/2018/05/Boko-Haram-Beyond-the-Headlines_Chapter-2.pdf. Accessed 18 June 2019

Chapter 3
Temporal Probabilistic Rules and Policy Computation Algorithms

The principal goal of this book is to develop a model that is capable of predicting attacks by Boko Haram – just as we have previously done in the case of Lashkar-e-Taiba (Subrahmanian et al. 2012) and the Indian Mujahideen (Subrahmanian et al. 2013). Chapter 1 of this book shows that the predictive models used are highly accurate in predicting many of the types of attacks that Boko Haram carries out. However, as stated in (Subrahmanian and Kumar 2017), a good predictive model must have three components:

- First, the predictive model must be *accurate* – it must make predictions that are correct most of the time. It is important to note that accuracy is measured in machine learning through a variety of technical metrics. These include

  - Precision/Confidence: Of all predictions that an attack will happen, precision is the percentage of times the attack did in fact happen during the time frame of interest.
  - Recall: Of all attacks that did occur during the time frame of interest, recall refers to the percentage of those attacks that were predicted to occur.
  - F1-Score: There are many applications where the precision is very high and the recall is very low. The F1-score combines the previous two metrics by computing their harmonic mean.
  - Accuracy: The technical definition of accuracy simply computes the percentage of predictions made (attack will occur, and attack will not occur) that agree with what actually occurred or did not occur during the time frame in question. As a metric, accuracy can be wildly inappropriate. For example, if an attack only occurs in 10 of 100 months, a predictor that predicts that that attack will never occur will have an accuracy of 90% – but would have 0% recall and would never predict a single attack.
  - Other metrics for measuring predictive accuracy include the popular Area Under a ROC Curve.

All of these metrics lie in the [0,1] interval with a higher number denoting better predictive performance.

- Second, the predictive model's predictions must be *explainable* to a person who is knowledgeable in the field in which the prediction is being made. In the case of this book, that means that the predictive models' predictions must be explainable to a counterterrorism or a military or intelligence official. Such individuals are well trained in their field but are usually not experts in computer science. The Temporal Probabilistic (TP) rule paradigm used in this book and described in more detail in this chapter focuses on explainability. TP-rules were first introduced in (Subrahmanian et al. 2012) and build upon prior work merging probabilities and time together within rules (Dekhtyar et al. 1999).
- Third, the predictive model must provide enough insight to enable *actionability.* An expert on Boko Haram not only needs to know what a predictive model of Boko Haram predicts they might do in the coming six months; it also needs to provide suggestions on how Boko Haram's attacks might be mitigated. Such suggestions of strategic actions will enable policy makers to shape actions on the ground that reduce the intensity of Boko Haram's attacks. We will discuss a policy computation algorithm later in this chapter.

3.1 Boko Haram Data

Our Boko Haram dataset is represented as a relational table – for non-computer scientists, the tables are just spreadsheets. A *row* in the Boko Haram dataset corresponds to a month during the period 2009–2016 (8 years in all). Though "8 years of data" may sound like a lot to some, for computer scientists, this dataset consisting of 96 rows is in fact very small. The columns in the Boko Haram dataset consist of two types of variables:

- *Dependent (Attack) Variables:* These are the types of attacks we want to predict. In this book, we look at predicting Boko Haram attacks 1, 2, 3, 4, 5, 6 months into the future. In other words, this book provides methods to predict – say on Jan 1, 2021 – the attacks that will happen in Jan 2021, sometime in the Jan 2021 – Feb 2021 period, and so forth until and including the Jan 2021 – June 2021 period. The same techniques described in this section can be easily adapted to make such predictions in other periods of time (say for the next year).
- *Independent (Environmental) Variables:* These variables capture various aspects of the context in which Boko Haram operates. They include information that falls into three broad categories:

  - *Actions that Boko Haram takes.* Some actions that Boko Haram takes such as addressing the Nigerian government directly in public communications, forcibly recruiting personnel to their cause, and use of child soldiers are examples of actions taken by Boko Haram.

- *Social, Cultural, Political, Economic, Military Variables.* These constitute the bulk of the independent variables. These variables may include, for instance, Boko Haram actively pushing out messages about the religious agenda, variables linked to elections in Nigeria, variables related to the availability of jobs, treatment of women, and more. They may also include actions taken by third parties that affect the environment in which Boko Haram operates – such as raids, arrests and kills of Boko Haram personnel by Nigerian security forces, international trials or condemnation of Boko Haram, and more.
- *Group-related Internal Variables.* Such variables include structural information about the group such as the nature of the leadership of the group, whether there is internal dissension within the group, and the nature of the group's relationships with other armed groups.

Appendix B contains a comprehensive summary of the variables used in our Boko Haram dataset.

3.2 Temporal Probabilistic Rules

At a high level, a temporal probabilistic rule is an expression of the form "If an environmental condition C holds during month m, then an attack A will (or will not) occur in month $m + \delta$". Thus, δ is a time delay which specifies how many months in the future, the attack A will occur.

An environmental condition is a condition over the environmental variables. In the case of the Boko Haram data, all variables are binary (yes/no). If E is an environmental variable, then $E = 0$ and $E = 1$ are *environmental atoms*. Similarly, if A is an attack variable, then $A = 0$ and $A = 1$ are *attack atoms*.

A TP-rule r is an expression of the form

$$EA_1 \& \ldots \& EA_n \rightarrow AA : \delta$$

where $EA_1, \ldots, EA_n$ are environmental atoms and AA are attack atoms, and $\delta \geq 1$ is an integer. The integer δ is the time delay mentioned above – in this book, we restrict δ to be between 1 and 6 as all predictions are made for the next 6-month period.

Despite the apparently formal statement of the definition of TP-rules given above, the rule can be easy read in English as: "*if* $EA_1, \ldots, EA_n$ *are all true in a given month*, then AA will (or will not) occur in month $m + \delta$". The rule is read as "attack A will occur" when the attack atom AA has the form $A = 1$ for some attack A and "attack A will not occur" when the attack atom AA has the form $A = 0$.

We call $EA_1 \& \ldots \& EA_n$ the *pre-condition* of the above rule and AA the *conclusion* of the above rule.

For example, consider a TP-rule about Boko Haram says that: "if Boko Haram reportedly has no foreign members and Boko Haram members have reportedly been imprisoned by the government during a month m, then Boko Haram will carry out sexual violence 5 months later". In this case,

- EA_1 is the atom "Boko Haram reportedly has no foreign members" and
- EA_2 is the atom "Boko Haram members have reportedly been imprisoned by the government" and
- AA is *Sexual Violence* = 1 and
- $\delta = 5$.

We call such rules *positive* TP-rules because they predict an attack will occur, i.e. the attack atom AA in the conclusion of such rules are of the form $A = 1$.

Sometimes, we are interested in predicting that an attack will not occur, i.e. when the attack atom AA in the conclusion of such rules are of the form $A = 0$. We call such rules *negative rules*.

An example of a negative rule is the rule "if Boko Haram was reportedly not recruiting/training and/or deploying individuals of a young age and Boko Haram was not designated as an international terrorist organization during a month m, then Boko Haram will not carry out arson attack one month later". In the case of this rule:

- EA_1 is the atom "Boko Haram was reportedly not recruiting/training and/or deploying individuals of a young age" and
- EA_2 is the atom "Boko Haram was not designated as an international terrorist organization" and
- AA is *Arson* = 0 and
- $\delta = 1$.

Our temporal-probabilistic rule mining framework can easily extract such rules from data of the kind described in Sect. 3.1.

Every TP-rule EA_1 & ... & $EA_n \rightarrow AA : \delta$ has 5 associated statistics that we compute.

Confidence The confidence of the above rule is the conditional probability of AA being true in month $(m + \delta)$, given that the precondition is true in month m. We want TP-rules to have high confidence.

Negative Confidence The negative confidence of the above rule is the conditional probability of AA being true in month $(m + \delta)$, given that the precondition is false in month m. We want TP-rules to have low negative confidence. Intuitively, when a TP-rule has high confidence and low negative confidence, then the precondition EA_1 & ... & EA_n serves as a beacon. When it is true (turned on), it predicts that AA will be true δ months in the future. When it is false (turned off), it predicts that AA will not be true δ months in the future. The bigger the "spread" (difference) between confidence and negative confidence, the better the rule distinguishes between whether AA is true δ months into the future and when AA is false δ months into the future.

Inverse Confidence The inverse confidence of the above rule is the conditional probability of the precondition of the above rule being true in month $(m - \delta)$, given that AA is true in month m. We want TP-rules to have high inverse confidence.

Lift Sometimes a rule can have high confidence and inverse confidence because AA is almost always true – in this case, the precondition is not of much use because AA is true regardless of whether the precondition is true or not. The *lift* of a rule is the ratio of the confidence of the rule, divided by the prior unconditional probability of the conclusion (AA) of the rule. We want a lift to be greater than 1 – the larger the better.

Support The support of the rule is the percentage of months m such that EA_1 & ... & EA_n is true in month m and AA is true in month $(m + \delta)$. We want support to also be high, but at least 5% if possible. Intuitively, the support of a rule makes sure that there are sufficiently many cases in which EA_1 & ... & EA_n and AA hold in the appropriate months – that the rule is not a "one of" type of situation.

All of the above metrics lie between 0 and 1 except for lift. We want high values for all of the above metrics except for negative confidence (which should be low) – and in the case of lift, the value should be higher than 1.

TP-rules build upon the idea of a class of rules called *association rules* (Agrawal and Srikant 1994) which were first used to help companies like Walmart perform *market basket* analysis which seeks to understand customer buying patterns in grocery stores, department stores and more. For example, association rules can be used to identify the probability that customers who buy milk and eggs may also buy bread. Association rule *mining* algorithms try to identify rules which are guaranteed to satisfy thresholds for support and confidence. The best-known algorithms for association rule mining are the Apriori algorithm (Agrawal and Srikant 1994), but there are many more sophisticated implementations which are far more efficient (Tan et al. 2016). However, association rules cannot capture the temporal dependencies that are necessary for reasoning about terrorist groups such as Boko Haram. TP rules, first introduced in (Subrahmanian et al. 2012), can be viewed as a combination of temporal probabilistic logic programs (Dekhtyar et al. 1999) and association rules.

Mining TP-rules from a body of data can be done via specialized methods such as those in (Subrahmanian and Ernst 2009) or via a simple application of association rule mining methods. Suppose we assume (as is the case with the Boko Haram dataset) that each row in the input table T corresponds to a month m. Then to mine association rules whose conclusion is of the form $A = j : \delta$ for $j \in \{0, 1\}$ and for $\delta \in \{1, ..., 6\}$, we create a new table $T_{j, \delta}$ as follows.

(i) Delete all dependent variables except for the dependent variable A.
(ii) For the rows corresponding to month m in T s.t. $m \leq MAX - \delta$, replace the entry in column A by the value of A in month $(m + \delta)$. Here, MAX is the last month in the dataset.
(iii) Delete rows $(MAX - \delta + 1, ..., MAX)$.

(iv) For the row corresponding to month m in T, replace the entry in column A by the value of A in month $(m + \delta)$.

We can then apply any standard association rule mining algorithm to $T_{j,\delta}$ in order to get TP-rules with $A = j : \delta$ in the conclusion of the rule and with a delay of δ.

Because standard association rules typically support only confidence, support and (sometimes) lift thresholds, the resulting set of TP-rules needs to go through a further filtering step to make sure that the final set of rules returned satisfy user-specified support, confidence, inverse confidence, negative confidence, and lift requirements.

3.3 Policy Computation Algorithm

TP-rules provide great insight into the behavior of a terrorist group such as Boko Haram. However, predicting the types of attacks that Boko Haram might carry out in the future is not enough. Counter-terrorism strategists and decision makers would like to understand what they can do in order to mitigate attacks by Boko Haram and reduce the probability of such attacks happening. To see how this can be achieved, consider a very simple example of two positive TP rules (predicting some attacks). For example, consider

$$EA_1 \ \& \ EA_2 \rightarrow A_1 = 1,1$$

$$EA_1 \ \& \ EA_3 \rightarrow A_2 = 1,1$$

These two positive TP-rules predict two different types of attacks, each with a one-month delay. Suppose our counter-terrorism policy analyst wants to recommend a policy that has a good chance of preventing both these attacks. He can do so in many ways:

(i) He might suggest taking some actions that would cause EA_1 to become false. Because EA_1 is a pre-condition of both TP-rules shown above, making EA_1 false would kill two birds with one stone: it would falsify the pre-condition of both rules. But this means that the probability of A_1 would be below the Negative Confidence threshold used by the TP rule mining algorithm, i.e. it would have a low probability of occurring. The same reasoning would also ensure that A_2 would also have a low probability of occurring.

(ii) On the other hand, the policy analyst could recommend that steps be taken to cause EA_2 and EA_3 to *both* be falsified, i.e. steps be taken to ensure that $(\neg EA_2 \ \& \ \neg EA_3)$ be made true. In this case again, because EA_2 is false, the probability of A_1 would be low and because EA_3 is false, the probability of A_2 would also be low.

Of course, it is possible that falsifying an environmental condition such as EA_1 is simply out of the realm of feasibility. For instance, in the case of Boko Haram, asking them to give up their desire for a fundamentalist Islamic state that practices Sharia law may be infeasible, just as it might be infeasible to expect groups such as Lashkar-e-Taiba demanding that Kashmir join Pakistan might be infeasible.

A second complication is that even if it is possible to falsify each of EA_2 and EA_3 individually, falsifying the two of them simultaneously may be infeasible. This could be for many reasons such as cost, politics, and other factors.

We therefore assume the existence of a feasibility predicate ϕ which takes a set of environmental atoms and determines if it is feasible to jointly falsify them. Of course, if $\phi(X) = true$ & $Y \subseteq X$ then we should also have that $\phi(Y) = true$—otherwise the feasibility predicate would not make sense.

Given a set PTP of positive TP-rules (rules predicting attacks) and a feasibility predicate ϕ as input, our Policy Computation Algorithm now generates a policy to mitigate these attacks as follows.

(i) **Compute minimal hitting sets.** Given a set X of sets, a *hitting set* for X is any set Y such that Y intersects every set in X. Suppose Y is a hitting set w.r.t X. Y is a *minimal hitting set* w.r.t. X if there is no other hitting set $Y' \subset Y$ w.r.t. X. Algorithms to compute all minimal hitting sets of a given set X exist in the literature (Gaine-Dewar and Vera-Licona 2017). As a first step, we compute the set $MHS(PTP)$ of all minimal hitting sets of the set $\{Pre\_Cond(r) | r \in PTP\}$ where $Pre\_Cond(r)$ is the set of all pre-conditions of a TP-rule r.

(ii) **Feasibility Filtration.** We now delete from $MHS(PTP)$, all minimal hitting sets that do not satisfy the feasibility requirement. The result is the set $Feas\_MHS(PTP) = \{Y \in MHS(PTP) | \phi(Y) = true\}$.

(iii) **Final Answer.** The final set of policies is now given by $\{\neg Y | Y \in Feas\_MHS(PTP)\}$ where the negation of a set is the set consisting of the negation of each element in it, i.e. $\neg Y = \{\neg EA | EA \in Y\}$.

For instance, let us consider a slightly expanded version of the two TP-rule set above.

$$EA_1 \ \& \ EA_2 \rightarrow A_1 = 1:1$$

$$EA_1 \ \& \ EA_3 \rightarrow A_2 = 1:1$$

$$EA_1 \ \& \ EA_2 \rightarrow A_3 = 1:1$$

In this case, the first step of the Policy Computation Algorithm would find two hitting sets

$$H_1 = \{EA_1\}$$

$$H_2 = \{EA_2, EA_3\}$$

Thus, the set $MHS(PTP) = \{H_1, H_2\}$. Suppose the feasibility predicate says

$$\phi(X) = \begin{cases} false \text{ if } H_1 \subseteq X \\ true \text{ otherwise} \end{cases}$$

Intuitively, this says that falsifying EA_1 is not feasible and hence falsifying any set X which is a superset (or equal to $\{EA_1\}$) is also infeasible. Thus, at the end of the feasibility filtration step, we have $Feas\_MHS(PTP) = \{H_2\}$. Basically this says that in our small toy example consisting of just 3 TP rules, there is only one feasible policy which will be obtained from H_2. That policy is obtained by falsifying everything in H_2, i.e. the final policy is $\{\neg EA_2, \neg EA_3\}$.

Our Policy Computation Algorithm (PCA for short) generates these policies that can then be presented to analysts who may bring their expertise (in this case on Boko Haram and Nigeria) to the table and subject these policies to further analysis. They may, for instance, decide later that some policies just won't work (for reasons not present in the data available to the machine learning models), some are too risky and perhaps some will alienate other partners not considered by the model, and more. It is important to note that these are high level policies and they could be implemented using any number of underlying tactics – we do not study this in this book.

We applied the PCA algorithm listed above to all the positive TP-rules presented in this book and derived the five policies presented in Chap. 1.

3.4 Sample Predictions

The authors have been generating prediction reports on the first of every month since January 1, 2019 – approximately 18 prediction reports have been generated thus far. As an example, we present the sample Boko Haram prediction report generated on Dec 1, 2019 in Appendix C. This report makes predictions for the following periods: December 2019, December 2019 – January 2020, December 2019 – February 2020, and so on till December 2019 – May 2020. We evaluated these predictions on June 1, 2020 and found that every single prediction was correct – this is not always the case though.

For example, the sample report in Appendix C contains the following predictions relating to sexual violence.

Timeframe	Dec	Dec–Jan	Dec–Feb	Dec–Mar	Dec–Apr	Dec–May
Likelihood	0.921	0.783	0.792	0.998	0.991	0.996
Event occurred:	Yes	Yes	Yes	Yes	Yes	Yes

For every one of the time periods, the probability that a suicide bombing will occur is well over 70%, suggesting that each of these events is highly likely (we consider any prediction of 70% or more to be highly likely, any prediction between

60–70% to be somewhat likely, predictions below 30% to be highly unlikely, and predictions between 30–40% to be somewhat unlikely).

Let us consider the rule SV-4 discussed later in Chap. 4.

TP-Rule SV-4

Sexual violence is committed in months in which:

- 3 months, Members of Boko Haram were allegedly imprisoned by the national government.
- 3 months earlier, Boko Haram communications did not address the government.
- 3 months earlier, Boko Haram openly expressed its willingness to negotiate or denied refusing talks.

Support = 0.41
Probability = 67%, Inverse Probability = 82%, Negative Probability = 24% Lift = 1.7

This rule suggests that sexual violence occurs in months m (December 2019 in our case) when 3 months earlier (September 2019 in our case), three things were concurrently true. First, Boko Haram members were reportedly arrested in month $m - 3$. Our data shows that in September 2019, Boko Haram members were in fact imprisoned by the Government of Nigeria (Independent 2019). Second, our data did not identify any reports of communications by Boko Haram that addressed the Nigerian government. Third, our data shows no evidence that Boko Haram refused to participate in talks with the Nigerian Government in September 2019. When these 3 pre-conditions are simultaneously true in month $m - 3$, the TP-rule SV-4 shown above says that in 67% of the case, sexual violence occurs 3 months later. Moreover, when at least one of these three pre-conditions is false, then sexual violence occurs three months later with a probability of only 24% which is a steep drop. Moreover, the lift of 1.7 shows that there is a 70% increased chance of sexual violence occurring as opposed to the base probability of sexual violence occurring in a given month.

Caveat

It is important to note that the best predictor for predicting whether an event occurs next month (Dec 2019 in the example) may be different from the best predictor for whether an event occurs sometime in the next 2 months (Dec 2019 – Jan 2020 in the above example). As a consequence, the prediction probabilities can vary.

3.5 Conclusion

In this chapter, we have provided a quick overview of Temporal Probabilistic rules and our Policy Computation Algorithm. Because this book is intended for a broad audience, we do not go into the gory math and computational details of these approaches – but rather have limited ourselves to a brief overview and presentation of the ideas underlying the algorithm.

We conclude with a cautionary note: correlation is not causation. The TP-rules presented in subsequent chapters of this book are *not* causal rules. Rather, they are rules that explain the types of predictions shown in Chap. 1.

References

Agrawal R, Srikant R (1994, September) Fast algorithms for mining association rules. In: Proceedings of the 20th international conference on very large data bases, VLDB, vol 1215, pp 487–499

Dekhtyar A, Dekhtyar M, Subrahmanian VS (1999) Temporal probabilistic logic programs. In: Proceedings of 1999 international conference on logic programming, New Mexico, November 1999, pp 109–123

Gainer-Dewar A, Vera-Licona P (2017) The minimal hitting set generation problem: algorithms and computation. SIAM J Discret Math 31(1):63–100

Independent (2019) OPPI: army arrests undercover aids of Boko Haram. https://www.independent.ng/oppi-army-arrests-undercover-aides-of-boko-haram/. Accessed 8 June 2020

Subrahmanian VS, Ernst J (2009) Method and system for optimal data diagnosis. U.S. Patent 7,474,987. University of Maryland, Baltimore

Subrahmanian VS, Kumar S (2017) Predicting human behavior: the next Frontiers. Science 355(6324):489

Subrahmanian VS, Mannes A, Sliva A, Shakarian J, Dickerson J (2012) Computational analysis of terrorist groups: Lashkar-e-Taiba. Heidelberg, Germany, Springer

Subrahmanian VS, Mannes A, Roul A, Raghavan RK (2013) Indian Mujahideen: computational analysis and public policy. Springer, Cham

Tan PN, Steinbach M, Kumar V (2016) Introduction to data mining. Pearson Education India

Chapter 4
Sexual Violence

From 2009 to 2016, Boko Haram committed acts of sexual violence in forty-four of the eighty-nine months we studied. In the last two years (2015 and 2016) of the period we study in this book, sexual violence occurred in twenty-three of those twenty-four months. Figure 4.1 depicts Boko Haram's use of sexual violence over time. Boko Haram's campaign of sexual violence demonstrates their commitment to using heinous crimes to further their agenda.

Generally, Boko Haram uses sexual violence as one of its many terror tactics. The group uses rape, forced marriages, prostitution, as well as the sale of enslaved women and girls in its campaign of terror (Lederer 2015). Thousands of girls and women have been abducted by the group. They have endured horrific abuses and to add to their suffering they are often rejected by their families after their escape, release or rescue (Okafor 2017). It is estimated that the group has abducted over 4000 children from their schools (Okafor 2017). Some specific events in Boko Haram's campaign of sexual violence are:

- **2012**. Boko Haram spreads propaganda claiming they will kidnap "infidel women as slaves" (Maiangwa and Agbiboa 2014).
- **April 14, 2014**. Two hundred and seventy-six girls were abducted from a boarding school in Chibok, Borno State. The girls were taken at night during their final examinations (BBC News 2017).
- **May 6, 2014**. Eight girls were kidnapped by militants suspected to be members of Boko Haram. The girls' ages ranged from eight to fifteen and they were abducted from a village in Nigeria's Borno State (Mark 2014). On the same day, a video was released by the leader of Boko Haram threatening to sell the girls abducted in the raid on the boarding school in Chibok (The Guardian 2014).
- **May 16, 2016**. One of the girls taken from the Chibok School was found with her four-month-old baby and a suspected Boko Haram fighter (BBC News 2016).

V. Subrahmanian et al., *A Machine Learning Based Model of Boko Haram*,
Terrorism, Security, and Computation, https://doi.org/10.1007/978-3-030-60614-5_4

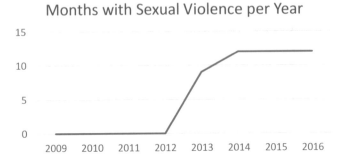

Fig. 4.1 Number of reported attacks involving sexual violence by Boko Haram

- **October 13, 2016**. Over two years after being abducted from their school in Chibok, 21 girls were released by Boko Haram. (Patience M 2016)
- **March 19, 2018**. Boko Haram kidnaps 110 girls from the Dapchi School in Dapchi, Nigeria. The youngest girls taken during the raid were eleven years old (Gopep et al. 2018).
- **April 15, 2018**. Four years after being abducted from their school in Chibok, 112 of the 276 kidnapped girls are still missing. The Nigerian Government claims they are negotiating for their release, while a Nigerian journalist claims only 15 of the 112 girls are still alive (BBC News 2018).
- **June 20, 2018**. The United Nations urges the government of Nigeria to provide aid and support to the women and children who face social isolation and rejection due to their abuse at the hands of Boko Haram (Aido 2018).
- **June 27, 2018**. Nigeria is ranked number nine on the list of worst countries to be a woman. Nigeria was also tied with Russia as the fourth most dangerous country for human trafficking (Reuters 2018).

Over the past ten years, Boko Haram has waged a campaign of sexual violence. Figure 4.2 shows the most commonly occurring independent variables linked to the TP-Rules we derived for predicting occurrence of sexual violence 4 months in advance.

Our research discovered the most rules when predicting sexual violence 4 months in advance. The 10 variables listed below are the ones that were most frequently present in TP-rules for predicting sexual violence by Boko Haram.

- *Members of Boko Haram were allegedly imprisoned by the national government.* If members of Boko Haram are imprisoned by the Nigerian Government and certain other conditions hold, then we can reliably predict that sexual violence will occur 5 months in the future using a TP-Rule we derived.
- *BH communications did not address the government.* Boko Haram can be predicted to commit sexual violence in the future after months in which they do not explicitly address the government in their communications and certain other conditions hold. This variable is often paired with other variables to generate several TP-Rules.

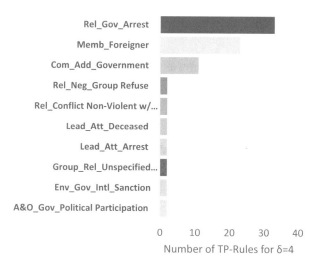

Fig. 4.2 Most frequently occurring variables for TP-rules that predict 4 months out in the future

- *BH did not reportedly have foreign members.* If there were no reports of members of the group being foreigners and certain other conditions hold, then sexual violence can reliably be predicted to occur five months in the future using a TP-Rule we derived.
- *BH openly declines to negotiate or agree to talks with the national government.* Whether Boko Haram agrees, or refuses, to negotiate with the government is an important factor in TP-Rules for predicting sexual violence. These rules predict whether sexual violence does or does not occur in future months.
- *BH was reportedly involved in non-violent inter-organizational conflict with another NSAG.* In months where Boko Haram is involved in a non-violent conflict with another NSAG and certain other conditions hold, we derived TP-Rules that predict the occurrence of sexual violence. This variable can also be used to predict a lack of sexual violence if there is no conflict.
- *BH reportedly denied the death of a prominent leader and/or the leader(s) is not dead as previously suspected.* In months where the leadership of the group is reported to be alive and certain other conditions hold, we derived TP-rules that reliably predict sexual violence 4 months in advance. This suggests that the leadership of Boko Haram may be linked to sexual violence though of course, the rules do not constitute a "smoking gun", just a consistent pattern.
- *The leadership of BH reportedly was not imprisoned.* Similar to the previous feature, when the leadership of Boko Haram is not imprisoned and certain other conditions hold, we derived TP-rules that predict sexual violence 4 months in advance.
- *BH did not reportedly give unspecified support to another NSAG.* In months where Boko Haram did not give aid to another group of non-state actors and

certain other conditions hold, we derived TP-rules predicting future sexual violence.

- *International sanction(s) on the government were reportedly not instated or revoked.* In months where international sanctions on the Nigerian Government were either removed or not applied and certain other conditions hold, we derived TP-rules that predict the occurrence of sexual violence 3 months in the future.
- *BH did not reportedly strive for unspecified political aspirations and objectives or the political participation of a specific person.* When Boko Haram is not looking to actively further their political agenda and certain other conditions hold, we derived TP-rules that predict sexual violence 4 months in the future.

It should be noted that the previously listed features are highly relevant in predicting the occurrence of sexual violence at different intervals in the future, not just 4 months. The following variable did not make the top ten most prevalent when making TP-Rules but is covered in this chapter.

- *The Absence of Explicitly Advocating for Religious Rule.* Acts of sexual violence can be linked to Boko Haram not actively advocating for religious rule in a given month. Using this variable, we were able to derive a TP-Rule to predict the occurrence of sexual violence.

We also discovered rules from the data predicting when Boko Haram would *not* carry out sexual violence.

- *The government did not reportedly shut down BH's offices.* Boko Haram not committing acts of sexual violence can be linked to the government not shutting down BH's offices 1–5 months prior and we derived TP-Rules using this variable.
- *BH reportedly did not, stopped or denied employing child soldiers.* A lack of sexual violence can be associated with Boko Haram reportedly not using child soldiers in a given month. Using this variable, we developed TP-Rules which predicts a lack of sexual violence 1–5 months in advance.
- *Military aid has been suspended or denied and BH reportedly did not, denies or stopped forcibly recruiting people.* In months where the Nigerian government is not receiving military aid and Boko Haram is not forcing new members into its ranks, we were able to derive a TP-Rule to predict a lack of sexual violence either 2 or 4 months in the future.
- *Military aid has been suspended or denied and BH reportedly did not, denies or stopped forcefully recruiting people and a third variable.* Using the same two variables in the preceding bullet with another variable, several TP-Rules can be derived to predict a lack of sexual violence 1, 3, and 5 months in the future. The third variable used will determine the future month in which sexual violence can be predicted to not occur. For example, if BH was not designated as an international terrorist organization in a month where the other two variables were satisfied, there is a link between those variables and an absence of sexual violence one month in the future.

- *Military aid has been suspended or denied and Foreign states/international institutions reportedly did not accuse BH of human rights abuses and a third independent variable.* Absence of sexual violence is associated with: military aid not going to the government, BH not being accused of human rights abuses, and either BH not using news media and periodicals to get their message out, or the government reportedly not accused of committing war crimes. We were able to generate TP-Rules using combinations of these variables to predict an absence of sexual violence by Boko Haram.
- *Military aid has been suspended or denied and a combination of several other variables.* After military aid has been denied or cut off from the government, and one or two more conditions are met, we derived TP-Rules to predict a lack of sexual violence 2–5 months in the future. The other variables paired with the suspension/denial of military aid are: the Nigerian government did not declare a State of Emergency, no direct negotiations between BH and the Government of Nigeria took place in that month, BH was not designated as an international terrorist organization in that month, as well as several others.

The variables examined in this study show how certain events, and combinations of events in the same month, are linked to whether Boko Haram will commit acts of sexual violence. The TP-Rules we derived as a result are discussed in the following sections.

4.1   Sexual Violence in the Absence of Seeking Religious Rule

In months where Boko Haram is not advocating for establishment of Sharia Law in Nigeria, we were able to generate a rule predicting the occurrence of sexual violence one month in the future. We recognize that Boko Haram is always aiming to establish a religious government and this variable was coded based on whether BH issued a statement advocating for religious rule in a given month. If no statements were made during a month then Boko Haram was considered to not be advocating for religious rule. For example, in March of 2013 Boko Haram was not actively promoting a religious rule. In April, sexual crimes were committed by the group when women were abducted by the group (Human Rights Watch 2014). TP-rule SV-1 (Sexual Violence) presents the results of generating the rule with a one-month offset.

TP-Rule SV-1
Sexual violence is committed in months in which:

- 1 month earlier, Boko Haram is not reported as actively advocating for a religious government.

  Support = 0.42
  Probability = 81%, Inverse Probability = 92%, Negative Probability = 11%, Lift = 1.5

In months where Boko Haram is not actively advocating for a religious government, the same rule also holds for 2, 3, 4, and 5 month offsets. In each case the support, probabilty, inverse probablity, negative probablity, and lift are very similar to those presernted in TP-rule SV-1.

4.2   Sexual Violence in Relation to Imprisoned and Foreign Members

Boko Haram's members often find themselves imprisoned by the Nigerian Government. In instances like these, their fellow members often bargain with the government in order to secure the release of their comrades. One of the bargaining chips used by Boko Haram is girls and women they have captured. For example, in October 2016 twenty-one girls taken from their school in Chibok were released. It was reported that 4 Boko Haram commanders were released as part of a deal with the Nigerian government (Patience 2016). While these commanders were imprisoned, the soon to be released girls were most likely exposed to the types of abuse that has made Boko Haram notorious. By looking at whether Boko Haram's members are free or imprisoned, as well as their nationality, we were able to derive TP-rule SV-2. The findings of the rule with a 5-month offset are presented below.

TP-Rule SV-2
Sexual violence is committed in months in which:

- 5 months earlier, Boko Haram reportedly has members imprisoned by the government.
- 5 months earlier, Boko Haram reportedly has no foreign members.

  Support = 0.45
  Probability = 67%, Inverse Probability = 86%, Negative Probability = 21% Lift = 1.7

4.3   Occurrence of Sexual Violence in Relation to Imprisoned, Foreign Members and Other Variables

The findings of this section build on the results presented in Sect. 4.2. We were able to derive several more rules predicting the occurrence of sexual violence using the variables discussed in Sect. 4.2. By using a third independent variable, rules for 1, 2, 3, and 4-month offsets were developed. In total, 19 different variables could be used with the original two variables. One real life example of these rules in action could be seen in July to September of 2014. During that September, Nigerian women Hamsatu and Halima were abducted by Boko Haram after their city was captured by the group: once captured, the women were used as sex slaves and endured the horrors of rape and forced marriage on a daily basis (Sieff 2016). Two months prior, Boko Haram had members imprisoned by the Nigerian Government, the group reportedly did not have foreign members, and BH was not reportedly funding any other groups. These conditions were the independent variables used to derive TP-Rule SV-3 below.

TP-Rule SV-3
Sexual violence is committed in months in which:

- 2 months earlier, Boko Haram reportedly had members imprisoned by the government.
- 2 months earlier, Boko Haram reportedly had no foreign members.
- 2 months earlier, Boko Haram did not reportedly give financial support to another non-state armed group.

> *Support = 0.43*
> *Probability = 67%, Inverse Probability = 86%, Negative Probability = 19% Lift = 1.76*

This rule's results are similar to those of other possible TP-Rules. Some other third variables include but are not limited to: the government reportedly not releasing members of the group, BH's leadership reportedly was not fractious or imprisoned, BH openly expressed its willingness to negotiate or denied refusing talks, and no mediated, indirect negotiations between BH and its state were taking place.

4.4 Occurrence of Sexual Violence when Members of BH Are Imprisoned, and BH's Communications Did Not Address the Government

Similar to Sects. 4.2 and 4.3, this section makes use of the variable measuring whether any of Boko Haram's members were reported to be incarcerated by the government. However, this section takes into account that BH is not directing its communications at the government. If both of these conditions, as well as one of several third conditions are true, then we are able to generate rules for predicting the occurrence of sexual violence. In July 2015, Boko Haram expressed its willingness to negotiate a prisoner exchange. The terrorist group would release kidnapped schoolgirls in return for a release of extremists. Leading up to the proposed negotiations, BH had escalated its campaign in the region in an attempt to gain a better position to negotiate from (Belfast Telegraph Online 2015). Three months earlier in April, members of BH were being held by the government, BH did not release public communications directed at the government and BH was not refusing to negotiate. These three conditions were involved in the following TP-Rule SV-4 that we derived:

TP-Rule SV-4
Sexual violence is committed in months in which:

- 3 months earlier, Members of Boko Haram were allegedly imprisoned by the national government.
- 3 months earlier, Boko Haram communications did not address the government.
- 3 months earlier, Boko Haram openly expressed its willingness to negotiate or denied refusing talks.

Support = 0.41
Probability = 67%, Inverse Probability = 82%, Negative Probability = 24% Lift = 1.7

SV-4 is comparable to all of the other combinations of variables. They all produced the same values for each metric. Some of the other variables used in combination with the original two are, but not limited to: international sanctions on the national government were reportedly not applied or revoked, BH was reportedly not involved in non-violent inter-organizational conflict with another NSAG, and BH's leadership reportedly was not imprisoned. Going forward, this chapter will transition to look at variables that generate rules for when Boko Haram is predicted to not commit acts of sexual violence.

4.5   Lack of Sexual Violence when Boko Haram's Offices Are Not Shut Down

TP-Rule SV-5 is the first TP-rule we derived which can be used to predict the absence of sexual violence in a given month. This rule says that when BH's offices were not closed by the government, Boko Haram could be predicted to not commit sexual violence one month later. Before March of 2013, Boko Haram's offices had largely been left open by the government – but after this date, we found the Nigerian Government had increased its efforts to close BH's offices. This escalation in office closures was mirrored by the escalation of sexual violence that Boko Haram is known for. SV-5 has a higher support score than any of the preceding rules predicting the occurrence of sexual violence.

TP-Rule SV-5

Sexual violence is not committed in months in which:

- 1 month earlier, the government did not reportedly shut down Boko Haram's offices.

  Support = 0.46
  Probability = 69%, Inverse Probability = 91%, Negative Probability = 13% Lift = 1.78

4.6   Lack of Sexual Violence when Boko Haram Is Not Using Child Soldiers

When we derived TP-Rule SV-6, the independent variable we considered looked at Boko Haram's use of child soldiers. Boko Haram's use of child soldiers appears to mirror its use of sexual violence. From 2009 to 2013 we found that Boko Haram employed child soldiers in only one month. From 2013 onward, the group's use of child soldiers becomes much more prevalent and is almost constant during 2015 and 2016. This rise in the use of child soldiers follows the rise in use of sexual violence form 2013 onward. This rule is another example of how Boko Haram's relationship with children and sexual violence are intertwined. The findings of TP-Rule SV-6 are shown below. It should be noted that this rule has the largest support score recorded for a sexual violence rule.

TP-Rule SV-6

Sexual violence is not committed in months in which:

- 1 month earlier, Boko Haram reportedly did not, stopped or denied employing child soldiers.

  Support = 0.48
  Probability = 72%, Inverse Probability = 96%, Negative Probability = 7% Lift = 1.78

4.7   Lack of Sexual Violence when Military Aid Has Been Suspended or Denied and BH Is Not Forcefully Recruiting People

TP-Rule SV-7 was derived using two independent variables. If the Nigerian government's military aid has been suspended or denied, and Boko Haram is recruiting new members by force in the same month, then 4 months later we can reliably predict a lack of sexual violence using our rule. For example, consider the period of time from May 2012 to January 2013. From May to August of 2012, the Nigerian Government was not receiving military aid and BH was not recruiting by force. For the corresponding next four months, BH reportedly did not commit any acts of sexual violence. This rule is tied with rule SV-6 for the greatest support score. This rule is also exactly the same for predictions 2 months in the future. The result of TP-Rule SV-7 are presented below.

TP-Rule SV-7
Sexual violence is not committed in months in which:

- 4 months earlier, Military aid had been suspended or denied.
- 4 months earlier, Boko Haram reportedly did not, denies or stopped to forcefully recruit people.

  Support = 0.43
  Probability = 65%, *Inverse Probability* = 86%, *Negative Probability* = 20% Lift* = 1.74

4.8 Lack of Sexual Violence when: Military Aid Has Been Suspended or Denied, BH Is Not Forcefully Recruiting People, and the Satisfaction of a Third Variable

Like TP-Rule SV-7, TP-Rule SV-8 examines months when the Nigerian Government is denied military aid and Boko Haram is not reportedly recruiting people forcibly. If the government has been denied aid, BH is not recruiting by force, and a third variable are all satisfied concurrently in a month, then one month later we can use TP-Rules to predict sexual violence will not occur. The third variables that satisfy this rule include but are not limited to: BH was not designated as an international terrorist organization, no direct negotiations between BH and its state took place, and BH is not using news media and periodicals to get their message out. The choice of the third variable determines the month offset for the prediction. Depending on the variable, we can predict an absence of sexual violence 1, 3, or 5 months into the future. The results of one combination of variables is presented in TP-Rule SV-8. For example, in February 2010 all three conditions used in SV-8 were satisfied. Then three months later BH reportedly did not commit any sexual violence. This combination allows for predictions 3 months in advance.

TP-Rule SV-8
Sexual violence is not committed in months in which:

- 3 months earlier, Military aid has been suspended or denied.
- 3 months earlier, Boko Haram reportedly did not, denies or stopped to forcefully recruit people.
- 3 months earlier, no direct negotiations between Boko Haram and its state took place.

> *Support* = 0.40
> *Probability* = 67%, *Inverse Probability* = 81%, *Negative Probability* = 23% *Lift* = 1.72

4.9   Lack of Sexual Violence When: Military Aid Has Been Suspended or Denied, Foreign States/Institutions Do Not Accuse BH of Human Rights Abuses, and a Third Variable

In this section, denial of military aid to the Nigerian Government is once again one of the components of this TP-Rule. The second variable requires that Boko Haram is not accused of human rights abuses, while the third variable can be any one of a variety of conditions. Using these conditions, we automatically derive TP-Rules to predict an absence of sexual violence 2, 3, 4 and 5 months in the future depending on the third variable. The third variables can be but are not limited to: BH was not designated as an international terrorist organization, foreign state(s)/international organization(s) reportedly did not seek, deny or revoke prior allegations of war crimes against BH, and when no direct negotiations between BH and its state took place. The TP-Rule SV-9 below displays the results for one combination of variables.

TP-Rule SV-9
Sexual violence is not committed in months in which:

- 2 months earlier, Military aid has been suspended or denied.
- 2 months earlier, Foreign state(s)/ international institution(s) reportedly did not accuse Boko Haram of human rights abuses.
- 2 months earlier, Boko Haram did not issue messages addressing a justification of violence.

> *Support* = 0.42
> *Probability* = 66%, *Inverse Probability* = 84%, *Negative Probability* = 22%, *Lift* = 1.76

In real life, SV-9 can be seen in action a couple months after January 2011. During that January, the Government of Nigeria was not receiving military aid in the form of equipment or funds to directly fight BH, no institution had accused BH of human rights abuses and Boko Haram did not try to justify any of their violence. Two months later, reportedly, no sexual violence was attributed the group. In addition to SV-7, SV-8, and SV-9, the denial of military aid can be paired with many other variables to derive several other rules for predicting a lack of sexual violence. In total, 11 other rules could be derived.

4.10 Conclusions

For the past decade Boko Haram has waged a terror campaign with sexual violence being a trademark strategy. Abductions, forced marriages, human trafficking, and rape make up their horrendous tool kit. Our work has found several conditions that correspond to occurrences, or lack of sexual violence.

- Sexual violence appears to occur in months where the month before Boko Haram is not explicitly advocating for religious rule.
- Sexual violence seems to occur five months after Boko Haram has members imprisoned by the government and reportedly there are no foreign members amongst its ranks.
- Similar to the previous point, when both of the above conditions, as well as a third condition, are met, then sexual violence appears to follow 1–4 months in the future depending on the third variable.
- When Boko Haram's members are imprisoned, BH's communications are not being directed at the government and another third condition is satisfied, sexual violence often follows 3, or 4 months in the future.

The relationships between these conditions and sexual violence are not causative. However, they give us insight into the conditions where sexual violence is most prevalent. For example, when members of Boko Haram are imprisoned, security forces should anticipate a spike in sexual violence. The same could be said about when BH goes quiet about advocating for an Islamic government. In addition to finding conditions where sexual violence is prevalent, we were able to find conditions where we could predict an absence of sexual violence.

- Sexual violence seems to not occur in months where Boko Haram's offices were not shut down during the preceding month.
- When Boko Haram is reportedly not using child soldiers, then acts of sexual violence are usually not committed the next month.
- Sexual violence appears to not occur 2 or 4 months after the Nigerian government has been denied military aid and BH is reportedly not recruiting new members through force.

- Sexual violence appears not to occur 1, 3, and 5 months in the future when the above two conditions and one of several third conditions are true.
- In a month where Nigerian government has been denied military aid, Foreign states reportedly did not accuse BH of human rights abuses, and a third variable are satisfied, sexual violence appears not to occur 2–5 months in the future.

Just as the conditions related to the occurrence of sexual violence, these events are not causative in terms of their relationship with acts of sexual violence, but they do provide insight into the conditions where Boko Haram does not commit these heinous crimes. The theme appears to be that BH does not commit sexual violence when the Nigerian government is not receiving outside help. Knowing when sexual violence is less likely to occur would be a valuable tool for security forces fighting BH. By knowing when sexual violence is less likely to happen, resources could be directed to other missions in the fight against BH. Overall, having an idea of when sexual violence will or will not occur greatly aids security forces in a fight against an enemy that often decides how and when it engages.

4.11   Predictive Model/Reports Results

Figure 4.3 below shows the performance of our predictive model and predictive reports during 2019. From January 2019 to the writing of this book, our team has produced reports containing our predictions on the likelihood of several different attacks out by Boko Haram. Our predictive model uses over 90 months' worth of data collected for our research. The dataset is updated at the end of every month to reflect new occurrences in the Boko Haram conflict. We used 6 classifiers on our dataset: SVM, KNN, Random Forrest, Gaussian Naïve Bayes, Multinomial Naïve Bayes, and Logistic Regression. The model aims to predict whether sexual violence will, or will not, happen within a given timeframe. For example, if the offset is 2 then the model predicts whether or not an event will occur anytime during the next two months. The Fig. 4.3 below depicts the results of our predictions compared to the ground truth observed after the predictions we made.

Sexual Violence						
Time Period	1	2	3	4	5	6
Recall	83%	100%	100%	100%	100%	100%
Precision	56%	73%	82%	91%	100%	100%
Accuracy	55%	73%	82%	91%	100%	100%
F1	0.67	0.84	0.90	0.95	1.00	1.00

Fig. 4.3 Sexual violence results for our 2019 predictive reports

References

Aido S (2018) UN tasks Nigeria on stigmatization of 'Boko Haram wives, children'. Punch. https://punchng.com/un-tasks-nigeria-on-stigmatisation-of-boko-haram-wives-children/. Accessed 27 Nov 2018

BBC News (2016) Chibok girls: Kidnapped schoolgirl found in Nigeria. https://www.bbc.com/news/world-africa-36321249. Accessed 29 Nov

BBC News (2017) Nigeria Chibok abductions: what we know. https://www.bbc.com/news/world-africa-32299943. Accessed 28 Nov 2018

BBC News (2018) Chibok girls: many abductees dead, says journalist. https://www.bbc.com/news/world-africa-43767490. Accessed 28 Nov 2018

Belfast Telegraph Online (2015) Boko Haram offers kidnap girls swap. https://www.belfast-telegraph.co.uk/news/world-news/boko-haram-offers-kidnap-girls-swap-31362061.html. Accessed 6 Dec 2018

Gopep J, Searcey D, Akinwotu E (2018) Boko Haram's seizure of 110 girls Taunts Nigeria, and its leader. The New York Times. https://www.nytimes.com/2018/03/18/world/africa/boko-haram-dapchi-girls-nigeria.html. Accessed 27 Nov 2018

Human Rights Watch (2014) "Those Terrible Weeks in Their Camp" Boko Haram Violence against Women and Girls in Northeast. https://www.hrw.org/report/2014/10/ 27/those-terrible-weeks-their-camp/bookharam-violence-against-women-and-girls. Accessed 5 Dec 2018

Lederer E (2015) Extremists using rape, sexual slavery and forced marriage as terror tactics: UN report. The Globe and Mail. https://www.theglobeandmail.com/news/world/extremists-using-rape-sexual-slavery-and-forced-marriage-as-terror-tactics-un-report/article23940104/. Accessed 27 Nov 2018

Maiangwa B, Agbiboa D (2014) Why Boko Haram Kidnaps Women and Young Girls in North-eastern Nigeria. Conflict Trends. https://www.researchgate.net/profile/Daniel_Agbiboa/publication/266798115_WHY_BOKO_HARAM_KIDNAPS_WOMEN_AND_YOUNG_GIRLS_IN_NORTH-EASTERN_   NIGERIA/links/543cb03c0cf2c432f7421426.pdf. Accessed 29 Nov 2018

Mark M (2014) Suspected Boko Haram gunmen kidnap eight girls from village in Nigeria. The Guardian. https://www.theguardian.com/world/2014/may/06/suspected-boko-haram-gunmen-kidnap-girls-village-nigeria. Accessed 28 Nov 2018

Okafor J (2017) Boko Haram Continues to Brutalize Children – UN. Daily Trust. http://www-lexisnexis-com.dartmouth.idm.oclc.org/hottopics/lnacademic/. Accessed 27 Nov 2018

Patience M (2016) How did Nigeria secure the 21 Chibok girls' release from Boko Haram?. BBC News. https://www.bbc.com/news/world-africa-37667915. Accessed 28 Nov 2018

Reuters (2018) Factbox: which are the World's 10 most dangerous countries for women?. https://www.reuters.com/article/us-women-dangerous-poll-factbox/factbox-which-are-the-worlds-10-most-dangerous-countries-for-women-idUSKBN1JM01Z. Accessed 27 Nov 2018

Sieff K (2016) They were freed from Boko Haram's rape camps. But their nightmare isn't over. The Washington Post. https://www.washingtonpost.com/world/africa/they-were-freed-from-boko-harams-rape-camps-but-their-nightmare-isnt-over/2016/04/03/dbf2aab0-e54f-11e5-a9ce-681055c7a05f_story.html. Accessed 6 Dec 2018

The Guardian (2014) Boko Haram Leader: "We will sell the girls on the market" – Video. https://www.theguardian.com/world/video/2014/may/06/boko-haram-sell-girls-market-video. Accessed 28 Nov 2018

Chapter 5
Suicide Bombings

Since 2011, Boko Haram has consistently used suicide bombings to terrorize Nigeria. For instance, Boko Haram carried out suicide bombings in 38 of the 72 months during the 2011–2016 time period. The greatest concentration of suicide bombings occurred during 2015 and 2016. Boko Haram carried out suicide bombings in twenty-two of those twenty-four months. Below, Fig. 5.1 examines Boko Haram's escalating use of suicide bombings.

Figure 5.1 shows how as time has gone on Boko Haram's use of suicide bombing has increased over the years.

Another horrifying aspect of Boko Haram's suicide bombing campaign is the group's use of women and children to carry out the bombings. From 2011 to 2017, 338 suicide bombers were identified by their gender; 244 of those attackers were women (Kriel 2017). The use of children is also very prevalent. In the same time period, in 134 cases where the age of a bomber could be determined, 81 of the attackers were children or teenagers. Children and women are used by Boko Haram because they often raise less suspicion than adult men (Kriel 2017). The use of children in their prolonged suicide bombing campaign demonstrates the lengths Boko Haram will go to wage their terror campaign. Some instances of Boko Haram suicide bombings are listed below.

- **August 26, 2011**. A Boko Haram insurgent detonated a car bomb outside the Nigerian headquarters of the United Nations. Eighteen were killed along with the bomber (Murray and Nossiter 2011).
- **December 25, 2011**. Boko Haram suicide bombers set off multiple bombs on Christmas Day. Twenty-five people were killed when a church was bombed (The Telegraph 2011).
- **January 21, 2012**. In a series of bombings, 140 people were killed. In one instance, a suicide bomber detonated a carload of explosives in a regional police headquarters (Batty and Mark 2012).

V. Subrahmanian et al., *A Machine Learning Based Model of Boko Haram*,
Terrorism, Security, and Computation, https://doi.org/10.1007/978-3-030-60614-5_5

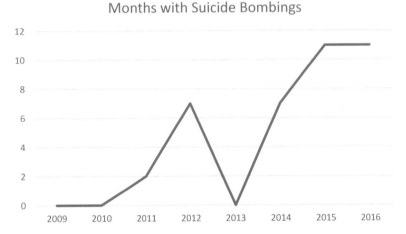

Fig. 5.1 Suicide bombings by Boko Haram from 2009–2016

- **April 26, 2012**. Boko Haram suicide bombers attacked the Nigerian newspaper "This Day" – six people were killed as a result of the blasts (Lobel 2012).
- **June 8, 2014**. A woman suicide bomber attacked a military barracks resulting in the death of a policeman (Chothia 2014).
- **November 10, 2014**. A boarding school was attacked by a Boko Haram suicide bomber; almost 50 boys were killed. Their ages ranged from ten to twenty years old (Nossiter 2014).
- **January 10, 2015**. A suicide bombing occurred in a market in Maiduguri killing twenty. The bomber was reportedly a girl estimated to be just ten years old (Nossiter 2015).
- **February 11, 2016**. A camp built to shelter people displaced by Boko Haram was bombed by two female attackers. Fifty-eight people were killed in the attack (CNN 2016).
- **November 21, 2017**. A mosque was bombed during morning prayer resulting in fifty being killed (Maclean 2017).
- **February 17, 2018**. A triple suicide bombing killed twenty in the Nigerian city of Maiduguri (Searcey 2018).

Boko Haram's campaign of suicide bombings has devastated communities in Nigeria for years. This chapter focuses on deriving TP-Rules to predict the occurrence or non-occurrence of suicide bombings. We briefly discuss some of the factors linked to predicting suicide bombings by Boko Haram. Figure 5.2 in the next page displays the most important features for predicting suicide bombings four months in advance.

- *The Absence of Explicitly Advocating for Religious Rule.* A key variable was months where Boko Haram was not explicitly promoting religious rule or a religious government. Though Boko Haram is consistently interested in promoting

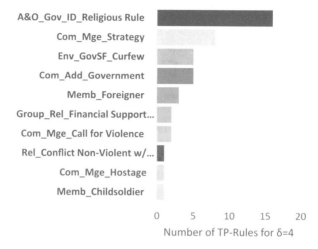

Fig. 5.2 Most Important Features Linked to Suicide Bombings

religious rule and sharia law, their public statements about it are frequent in some months and infrequent and/or absent in others. Using this variable, in combination with others, we were able to derive several TP-Rules to predict suicide bombings.

- *Boko Haram did not reportedly communicate their strategy.* We also looked at whether Boko Haram reportedly released communications detailing their tactics, or strategy. Using this variable in combination with others, we were able to derive TP-Rules to predict suicide bombings one to four months in advance.
- *A curfew was not enforced, or implemented, by government security forces.* When a curfew was not being enforced by security forces, and several other conditions were met, we were able to derive TP-Rules to predict suicide bombings one to four months in the future.
- *Communications released by Boko Haram did not address the government.* When Boko Haram's communications did not address the Nigerian government and other variables were satisfied in the same month, suicide bombings could be predicted four months in advance using a TP-Rule we derived.
- *Boko Haram reportedly did not have foreign members.* If Boko Haram reportedly has no foreign members in a given month, and some other conditions are true, then using a TP-Rule we derived we were able to reliably predict suicide bombings one month ahead of time.
- *Boko Haram provided financial support to another NSAG.* In a month where Boko Haram did not reportedly give financial aid to another group, and several other conditions were met, we were able to derive TP-Rules to predict the occurrence of suicide bombings.

- *Boko Haram is not releasing communications or propaganda calling for violence.* In months where Boko Haram was not releasing propaganda calling for violence, and other variables were satisfied, we were able to derive TP-Rules to predict the occurrence of suicide bombings one and two months in advance.
- *Boko Haram was not involved in a non-violent conflict with another NSAG.* When Boko Haram is not reportedly engaged in a non-violent conflict with another NSAG, and other conditions are met, we derived a TP-Rule to predict suicide bombings one and two months in advance.
- *Boko Haram's communications or propaganda contain messages about hostages taken by the group.* When Boko Haram does not release propaganda about hostages and other conditions are met in the same month, a TP-Rule we derived can be used to predict suicide bombings one and four months in advance.
- *Boko Haram recruits, trains and/or deploys individuals of young age.* The presence of child soldiers among Boko Haram is an important variable that appears – along with some other variables – in many TP-Rules. Using these rules, we can reliably predict both the occurrence, or absence, of suicide bombings one month in advance.

Below are some other variables used less frequently in the creation of TP-Rules that are linked to predicting the occurrence of suicide bombings:

- *Members of Boko Haram were reportedly arrested by the government.* A TP-Rule we derived made use of this variable along with several others. We were reliably able to predict suicide bombings three months after members of Boko Haram were arrested.
- *Negotiations between Boko Haram and the government reportedly not end with a cease-fire.* Using months where BH and Nigerian Government negotiations did not end with a cease-fire, we were able to derive a TP-Rule to predict suicide bombings one month in advance.
- *Boko Haram's members are reportedly not on trial.* Using a TP-Rule we derived, we were able to reliably predict suicide bombings one month in advance if Boko Haram's members were not on trial during a specific month.

The next set of variables were used to derive TP-Rules which enable us to predict an absence of suicide bombings in the future. These variables are often used in combination with others in order to create our TP-Rules.

- *Boko Haram reportedly did not, denied, or stopped creating YouTube Content.* When Boko Haram was not releasing propaganda on YouTube during a given month, and the group was not deploying child soldiers, we were able to derive a TP-Rule to predict a lack of suicide bombings one month later.
- *Boko Haram reportedly did not give unspecified support to another Non-State Armed Group (NSAG).* Some TP-Rules we derived depended on months where Boko Haram is not giving unspecified aid to other NSAG. When this condition and other variables were satisfied, we were able to predict an absence of suicide bombings three months in the future.

- *Boko Haram did not reportedly give military support to another NSAG.* Using a TP-Rule we derived, we are able to reliably predict a lack of suicide bombings one months after BH does not provide a military aid to a NSAG, and two other conditions are met.
- *No negotiations between Boko Haram, the respective state and/or mediator(s) were planned.* We derived a TP-Rule based on whether negotiations between the state, or a mediator, and BH were planned. If negotiations are not planned, and other conditions are met, then we are able to reliably predict a lack of suicide bombings one month in advance.

5.1 Suicide Bombings and Boko Harm's Use of Child Soldiers

In months where Boko Haram is recruiting and deploying child soldiers, we were able to derive a TP-Rule to predict the suicide bombings one month in the future. In recent years Boko Haram has turned to the abhorrent practice of using children to carry out their suicide attacks. BH uses children because they raise less suspicion among security forces and make attacking soft targets easier for the group (Kriel 2017). These children are often given money and promised entrance to heaven in return for detonating the bombs strapped to them (Kriel 2017). Our TP-Rule SB-1 (Suicide Bombing) is depicted below.

> **TP-Rule SB-1**
> Suicide Bombings occur in months in which:
>
> - 1 month earlier, Boko Haram recruits, trains and/or deploys individuals of young age.
>
>   *Support = 0.26*
>   *Probability = 79%, Inverse Probability = 61%, Negative Probability = 25% Lift = 1.78*

The results displayed in SB-1 are for a one-month offset. Using Boko Haram's use of child soldiers as the independent variable also allowed us to derive a TP-Rule with a four-month temporal offset as compared to the one-month temporal offset presented in TP-rule SB-1.

5.2   Suicide Bombings When Boko Haram Is Not Actively Promoting Religious Rule or Communicating Their Strategy

To derive TP-Rule SB-2, we examined the propaganda BH was releasing. In months when Boko Haram is not releasing propaganda to explicitly and actively promote religious rule, or share their strategy, we were able to use SB-2 to predict suicide bombings three months in advance. For example, in July of 2012, Boko Haram did not release any material (in the sources we checked) to promote an Islamic government, nor did the group publicly release details about their strategy. Three months later in October a suicide bombing occurred. The results of TP-Rule SB-2 are displayed below. It should be noted that this rule also holds true when predicting suicide bombings five months in advance.

TP-Rule SB-2
Suicide Bombings occur in months in which:

- 3 months earlier, Boko Haram was not actively striving for religious rule.
- 3 months earlier, Boko Haram did not reportedly communicate their strategy.

  Support = 0.32
  Probability = 65%, Inverse Probability = 74%, Negative Probability = 23% Lift = 1.74

5.3   Suicide Bombings in Relation to When Boko Haram's Members Are Arrested, and the Group Is Not Actively Promoting Religious Rule

TP-Rule SB-3 was derived using variables which looked at Boko Haram's propaganda and the status of the group's members. Using SB-3, we are able to predict suicide bombings three months after months in which Boko Haram did not actively promote religious rule and members of the group were arrested in the same month. For example, in March of 2011, members of Boko Haram were arrested and in addition, the group did not promote an Islamic government. Three months later in June of 2011, the headquarters of the Nigerian National Police was bombed when a suicide bomber drove a car bomb into the building (Pflanz 2011). The TP- rule we derived is shown below.

TP-Rule SB-3

Suicide Bombings occur in months in which:

- 3 months earlier, Boko Haram was not actively striving for religious rule.
- 3 months earlier, members of Boko Haram were reportedly arrested.

  Support = 0.32
  Probability = 65%, Inverse Probability = 74%, Negative Probability = 23% Lift = 1.74

5.4   Suicide Bombings in Relation to When Boko Haram's Members Are Arrested, the Group Not Actively Promoting Religious Rule and a Third Variable

Similar to SB-3, TP-Rule SB-4 looks at BH's religious propaganda and whether or not members of the group have been arrested. However, SB-4 also considers several additional variables. SB-4 was derived to allow us to make predictions about suicide bombings one month in advance. To make these predictions, Boko Haram must not actively promote religious rule, members of the group need to have been arrested, and a third variable all need to be satisfied in the same month. Some of these third variables include but are not limited to: Boko Haram is not using news media and periodicals to get their message out, no representatives of Boko Haram reportedly stood trial, Boko Haram was not designated as an international terrorist organization during that month, and Boko Haram communications did not address the public or call for violence or justification of violence or its strategy. The third variable used for SB4 below looks at Boko Haram's foreign members. For the variable to be satisfied, BH will need to reportedly not have foreign members in the given month.

TP-Rule SB-4

Suicide Bombings occur in months in which:

- 1 month earlier, Boko Haram was not actively striving for religious rule.
- 1 month earlier, members of Boko Haram were reportedly arrested.
- 1 month earlier, BH did not reportedly have foreign members.

  Support = 0.33
  Probability = 67%, Inverse Probability = 76%, Negative Probability = 20% Lift = 1.78

The above rule, SB-4, also holds true for predicting suicide bombings two months into the future. All of the same variables can be used as the third conditions and the results are similar to those above.

5.5   Suicide Bombings When Boko Haram's Members Are Not Trial, the Group Is Not Actively Promoting Religious Rule, and a Third Variable

When BH's members are not on trial, the group is not actively advocating for an Islamic government and a third variable are satisfied in the same month, we are able to use TP-Rules we derived to predict suicide bombings 1 month in advance. The third independent variables that can be used include but are not limited to: BH communications did not address the public or its hostages, BH was not designated as an international terrorist organization, and BH was not, denied, or stopped non-violent inter-organizational conflict with another NSAG. The TP-Rule SB-5 shown below uses when negotiations between BH and the government reportedly not end with a cease-fire as its third variable. The results of SB-5 are displayed below.

TP-Rule SB-5
Suicide Bombings occur in months in which:

- 1 month earlier, Boko Haram was not actively striving for religious rule.
- 1 month earlier, Boko Haram's members are reportedly not on trial.
- 1 month earlier, negotiations between Boko Haram and the government reportedly did not end with a cease-fire.

Support = 0.33
Probability = 67%, Inverse Probability = 76%, Negative Probability = 20% Lift = 1.78

An example of SB-5 can be seen in action in October of 2012 when all of the pre-conditions were satisfied. Then next month a suicide bombing occurred when a bus laden with explosives was driven into a church on a Nigerian Army barracks: 11 people were killed and 30 were injured (The Telegraph 2012).

In addition to the previously displayed TP-Rules, we were able to derive many other rules which can be used to predict future suicide bombings. We also derived many rules which can be used to predict an absence of suicide bombings. The following sections will transition to report the rules concerned with predicting a lack of suicide bombings.

5.6   A Lack of Suicide Bombings When Boko Haram Is Not Recruiting, Training and/or Deploying Children

As seen earlier in this chapter, Boko Haram often uses children as suicide bombers in order to wage their campaign of terror. This section covers a TP-Rule SB-6, a rule opposite to SB-1. Where SB-1 was used to predict suicide bombings, SB-6 can be used to predict their absence four months in the future. For example, in August 2013 Boko Haram was not reported as using children as combatants, nor was it recruiting them.[1] Four months later in December of that year no suicide bombings occurred. The results of TP-Rule SB-6 are displayed below.

> **TP-Rule SB-6**
> Suicide Bombings do not occur in months in which:
>
> - 4 months earlier, Boko Haram was not reportedly recruiting, training and/ or deploying individuals of a young age.
>
>   *Support* = 0.52
>
>   *Probability = 75%, Inverse Probability = 94%, Negative Probability = 12% Lift = 1.72*

It should be noted that this rule has a significantly larger support score compared to the rules previously discussed. In addition, this rule can also be used to predict an absence of suicide bombings five months in the future.

5.7   A Lack of Suicide Bombings When Boko Haram Is Not Recruiting, Training and/or Deploying Children Combined with another Variable

Building on the rule presented in Sect. 5.6, this section also looks at when BH is not using child soldiers. However, the rule presented in this action also utilizes a second variable. Possibilities for that second variable include but are not limited to: the government reportedly was not accused of committing war crimes during that month, international organizations reportedly did not seek, deny or revoke prior allegations of war crimes against BH, and personnel of the Nigerian security force(s) reportedly did not

---

[1] We note that there is a distinction between whether Boko Haram was *reported* to be using children and whether Boko Haram was *actually* using children as combatants in a given month. In practice, Boko Haram consistently uses children as combatants, but there are months where reports of their doing so are more common, e.g. in months when children were found at a Boko Haram location and/or during arrests and/or when child combatants detonated suicide bombs.

defect. The rule presented below, SB-7, is satisfied when BH is not reportedly deploying, recruiting, or training child soldiers, and the group reportedly did not, denies, or stopped creating content posted on YouTube.

TP-Rule SB-7
Suicide Bombings do not occur in months in which:

- 1 month earlier, Boko Haram was reportedly not recruiting, training and/or deploying individuals of a young age.
- 1 month earlier, Boko Haram reportedly did not, denied, or stopped creating YouTube content.

  Support = 0.49
  Probability = 77%, Inverse Probability = 86%, Negative Probability = 22% Lift = 1.78

TP-Rule SB-7 can also be used to predict an absence of suicide bombings two or three months in the future. In addition to the variables listed above, there are many others that can be paired with BH's absence of using child soldiers to derive TP-Rules. These other rules can be used to predict an absence of suicide bombings one, two, or three months in advance.

5.8   An Absence of Suicide Bombings When Boko Haram's Offices Have Not Been Closed in Combination with a Second Variable

TP-Rule SB-8 was derived by looking at when BH's offices were not closed by the government and when a second variable was satisfied. The possible second variables used by this rule include but are not limited to: the government was not reportedly accused of committing war crimes, and BH communications did not address the media. In SB-8 displayed below, the second variable was satisfied when BH reportedly did not give unspecified support to another NSAG.

TP-Rule SB-8
Suicide Bombings do not occur in months in which:

- 3 months earlier, The Nigerian government did not close down Boko Haram locations.
- 3 months earlier, Boko Haram reportedly did not give unspecified support to another NSAG.

  Support = 0.47
  Probability = 75%, Inverse Probability = 84%, Negative Probability = 25% Lift = 1.74

5.9 An Absence of Suicide Bombings When Boko Haram Is Not Reportedly Using Child Solders or Giving Military Support to Another Groups and No Negotiations with the Government Are Planned

TP-Rule SB-9, like several other rules, considers situations when Boko Haram was not reportedly using child soldiers as one of its pre-conditions. The other two conditions that need to be satisfied are: no negotiations between Boko Haram and the Government or mediator occurred, and BH did not reportedly give military support to other non-state armed groups (NSAGs). If all of these conditions are met, then we can use SB-9 to predict a lack of suicide bombings one month in advance. One example of this rule in action can be seen in December of 2009 when all of the conditions were satisfied. No suicide bombings occurred in January of 2010. The results of SB-9 are displayed below, and it should be noted that SB-9 can be used to predict a lack of suicide bombing two month in advance as well.

> **TP-Rule SB-9**
> Suicide Bombings do not occur in months in which:
>
> - 1 month earlier, Boko Haram was not reportedly employing child soldiers.
> - 1 month earlier, no negotiations between Boko Haram, the respective state and/or mediator(s) were planned.
> - 1 month earlier, Boko Haram did not reportedly give military support to another NSAG.
>
>   *Support* = 0.48
>   *Probability* = 77%, *Inverse Probability* = 84%, *Negative Probability* = 24% *Lift* = 1.78

5.10 Conclusions

Boko Haram's campaign of suicide bombings have inflicted death and terror on the people of Nigeria, and the surrounding countries, for years. Children, sometimes as young as ten, are either coerced or brainwashed into becoming weapons of terror for Boko Haram (Kriel 2017). Our work has found several conditions that can be linked to the group's use, or lack of use, of suicide bombers.

- Boko Haram's use of child soldiers seems to be closely related to their deployment of suicide bombers in the following month.

- When BH is not actively and explicitly advocating for religious rule and the group is not communicating their strategy or tactics, suicide bombings seem to follow three months later.
- When Boko Haram is not actively and explicitly promoting religious rule and members of the group have reportedly been arrested, suicide bombings usually occur three months in the future.
- When both of the two conditions in the previous bullet are satisfied along with a third condition, suicide bombings seem to follow in the next month.
- If Boko Haram's members are on trial, the group is not actively and explicitly promoting religious rule, and a third variable are all satisfied in the same month, then suicide bombings tend to occur one month later.

The relationships between these events and the suicide bombings that follow are not causative, but they provide a glimpse into the conditions linked to suicide bombings. For example, if Boko Haram abducts children and is reportedly training them, then security forces should anticipate new waves of suicide bombings. Similarly, military and police forces should expect suicide bombings after the members of BH have been arrested and the group has gone silent about promoting religious rule. Along with finding conditions linked to the occurrence of suicide bombings, our research found conditions related to a lack of attacks.

- If Boko Haram is reportedly not using or training child soldiers, then a lack of suicide bombings can be expected four months in the future.
- When Boko Haram is not using child soldiers and a second variable is satisfied, then suicide bombings are not expected to occur one, two, or three months in the future.
- If the Nigerian Government did not shutdown Boko Haram's offices and BH is not providing unspecified aid to another NSAG, then suicide bombings are not expected three months later.
- If in the same month Boko Haram is reportedly not using child soldiers, no negotiations with the state or a mediator are planned, and BH is not giving military aid to another NSAG, then following month is not expected to experience any suicide bombings.

Like the section discussing the circumstances related to the occurrence of suicide bombings, these relationships are not causative. But again, they provide a lens through which an environment can be evaluated for the likeliness of a suicide bombing in the coming months. Again, Boko Haram's use of children appears to be important in relationship to their suicide bombing campaign. If Boko Haram is not reportedly deploying children as combatants, then security forces could expect a lack of suicide bombings. Knowing the conditions closely related to Boko Haram's suicide attacks will be a helpful tool in combating their effectiveness. If security forces know when suicide bombings are most likely to occur, then commanders could better allocate their manpower and resources. Any edge given to security forces will help combat Boko Haram's relentless campaign of death and terror.

Suicide Bombings						
Time Period	1	2	3	4	5	6
Recall	89%	90%	82%	100%	100%	100%
Precision	89%	100%	100%	100%	100%	100%
Accuracy	82%	91%	82%	100%	100%	100%
F1	0.89	0.95	0.90	1.00	1.00	1.00

Fig. 5.3 Suicide bombings results for our 2019 predictive reports

5.11 Predictive Model/Reports Results

Figure 5.3 below shows the performance of our predictive model and predictive reports during 2019. From January 2019 to the writing of this book, our team has produced reports containing our predictions on the likelihood of several different attacks out by Boko Haram. Our predictive model uses over 90 months' worth of data collected for our research. The dataset is updated at the end of every month to reflect new occurrences in the Boko Haram conflict. We used 6 classifiers on our dataset: SVM, KNN, Random Forrest, Gaussian Naïve Bayes, Multinomial Naïve Bayes, and Logistic Regression. The model aims to predict whether suicide bombings will, or will not, happen within a given timeframe. For example, if the offset is 2 then the model predicts whether or not an event will occur anytime during the next two months. Figure 5.3 depicts the results of our predictions compared to the ground truth observed after the predictions we made.

References

Batty D, Mark M (2012) Scores killed in terrorist attacks in Nigeria. The Guardian. https://www.theguardian.com/world/2012/jan/21/nigeria-attacks-claimed-by-boko-haram. Accessed 4 Jan 2019

BBC News (2012) Nigerian suicide bomber targets Maiduguri mosque. https://www.bbc.com/news/world-africa-18834387. Accessed 3 Jan 2019

Chothia F (2014) Boko Haram crisis: Nigeria's female bombers strike. BBC News. https://www.bbc.com/news/world-africa-28657085. Accessed 4 Jan 2019

CNN (2016) Female suicide bombers kill 58 in a Nigerian camp meant to be a haven. https://www.cnn.com/2016/02/11/africa/nigeria-suicide-bombing-boko-haram/index.html. Accessed 4 Jan 2019

Kriel R (2017) Boko Haram favors women, children as suicide bombers, study reveals. CNN. https://www.cnn.com/2017/08/10/africa/boko-haram-women-children-suicide-bombers/index.html. Accessed 4 Jan 2019

Lobel M (2012) Nigeria's ThisDay newspaper hit by Abuja and Kaduna blasts. BBC News. https://www.bbc.com/news/av/world-africa-17861468/nigeria-s-thisday-newspaper-hit-by-abuja-and-kaduna-blasts. Accessed 4 Jan 2019

Maclean R (2017) Nigeria mosque attack: suicide bomber kills dozens. The Guardian. https://www.theguardian.com/world/2017/nov/21/nigeria-mosque-attack-teenage-suicide-bomber-kills-at-least-50. Accessed 4 Jan 2019

Murray S, Nossiter A (2011) Suicide bomber attacks U.N. building in Nigeria. The New York
    Times. https://www.nytimes.com/2011/08/27/world/africa/27nigeria.html. Accessed 4 Jan
    2019
Nossiter A (2014) Bomb at school in Nigeria kills nearly 50 boys. The New York Times. https://
    www.nytimes.com/2014/11/11/world/africa/nigeria-suicide-bomber-boko-haram.html.
    Accessed 4 Jan 2019
Nossiter A (2015) In Nigeria, New Boko Haram suicide bomber tactic: 'It's a Little Girl'. The
    New York Times. https://www.nytimes.com/2015/01/11/world/africa/suicide-bomber-hits-
    maiduguri-nigeria-market.html. Accessed 4 Jan 2019
Pflanz M (2011) Al-Qaeda-linked suicide bomber targets Nigeria police station. The Telegraph.
    https://www.telegraph.co.uk/news/worldnews/africaandindianocean/nigeria/8580438/
    Al-Qaeda-linked-suicide-bomber-targets-Nigeria-police-station.html. Accessed 2 Jan 2019
Searcey D (2018) Three suicide bombers kill at least 20 in Nigeria. The New York Times. https://
    www.nytimes.com/2018/02/17/world/africa/nigeria-suicide-bombing.html. Accessed 4 Jan
    2019
The Telegraph (2011) Boko Haram claims responsibility for Nigeria attacks. https://www.tele-
    graph.co.uk/news/worldnews/africaandindianocean/nigeria/8977493/Boko-Haram-claims-
    responsibility-for-Nigeria-attacks.html. Accessed 4 Jan 2019
The Telegraph (2012) 11 killed as Nigerian extremists breach military security to bomb church.
    https://www.telegraph.co.uk/news/worldnews/africaandindianocean/nigeria/9701745/11-
    killed-as-Nigerian-extremists-breach-military-security-to-bomb-church.html. Accessed 3 Jan
    2019

Chapter 6
Abductions

Boko Haram's campaign of abductions began in 2013 and picked up steam in 2014. In 2013 there were only two months with reported abductions, but in 2014 the number of months with reported abductions rose to ten. Between 2013 and 2016 Boko Haram reportedly abducted people in 27 of the 48 months. Figure 6.1 depicts the number of months with reported abductions for the time period we examined in this book. Overall, UNICEF reports that over 1000 children have been abducted by Boko Haram since 2013 (UNICEF 2018). Figure 6.1 shows how the number of abductions has generally increased over time.

Boko Haram abducts people for a variety of reasons. Boko Haram often targets young boys and girls since the group often brainwashes them into becoming suicide bombers; the girls also face the prospect of being used by the group as sex slaves and are forced to marry fighters in the group (Freeman 2018). The group also abducts people in order to hold them for ransom. In February of 2018, 110 girls were abducted from their school in Dapchi, Nigeria. In April of 2018, the girls were allegedly returned by Boko Haram "'unconditionally'" (Freeman 2018). However, it has been alleged that a ransom of $10 million was paid, following the precedent set when over $3 million was paid to release girls taken from the Chibok school in 2014 (Freeman 2018). These are some more recent developments in Boko Haram's campaign of abductions, the following instances of abductions further illustrate the group's crimes.

- **April 14, 2014**. 219 girls were abducted from their school in Chibok Nigeria during an exam. The abduction gained worldwide attention and spawned the #BringBackOurGirls social media campaign (BBC News 2017).
- **May 6, 2014**. Boko Haram threatens to sell the Chibok girls on the human trafficking market (Abubakar and Levs 2014).
- **March 2015**. An estimated 500 children were kidnapped by Boko Haram as military forces drove the group from the town of Damasak, Nigeria (Sieff 2017).

© The Author(s), under exclusive license to Springer Nature Switzerland AG 2021
V. Subrahmanian et al., *A Machine Learning Based Model of Boko Haram*,
Terrorism, Security, and Computation, https://doi.org/10.1007/978-3-030-60614-5_6

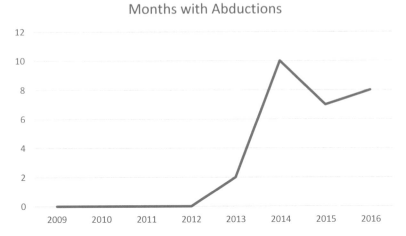

Fig. 6.1 Number of months per year when abductions occurred

- **May 18, 2016**. One of the abducted Chibok girls is found pregnant in a forest (The Guardian 2017).
- **August 2016**. It is estimated that 10,000 boys and teenagers have been abducted by Boko Haram since 2013 (PBS News Hour 2016).
- **October 13, 2016**. Boko Haram releases 21 of the Chibok girls after negotiating with the Nigerian government (The Guardian 2017).
- **June 2017**. Ten policewomen were abducted by Boko Haram after their convoy was ambushed, 69 people were also killed in the attack (Vanguard 2018).
- **July 2017**. Three lecturers from the University of Maiduguri were abducted during an oil expedition (Vanguard 2018).
- **February 10, 2018**. Ten policewomen and three lecturers from the University of Maiduguri were released by Boko Haram after negotiations with the Nigerian government (Vanguard 2018).
- **February 19, 2018**. A school in Dapchi, Nigeria was attacked by Boko Haram and 110 girls were kidnapped (BBC News 2018).
- **March 21, 2018**. 104 of the girls kidnapped form Dapchi are returned to their home by Boko Haram. Five of the girls perished in the ordeal, and one girl was still being held by the group for refusing to renounce Christianity for Islam (BBC News 2018).

Boko Haram's abductions have taken a terrible toll on the people of northeastern Nigeria for years. The focus of this chapter is to derive TP-Rules to predict when Boko Haram will abduct people, as well as when the group releases some of its prisoners. We now briefly discuss some of the events and conditions linked to Boko Haram's abductions. The variables used most frequently for developing TP-Rules with a two-month offset are displayed in Fig. 6.2. These variables and others are briefly discussed below.

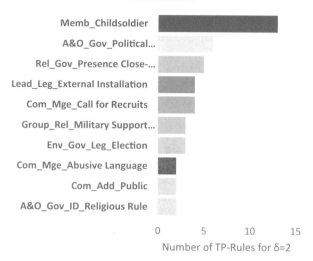

Fig. 6.2 Ten Main Factors Linked to Abductions by Boko Haram

- *Boko Haram recruits, trains and/or deploys individuals of young age.* The reported presence of child soldiers among Boko Haram's ranks is used, with other variables, to derive several TP-Rules. Using these rules, we can reliably predict both abductions and the release of prisoners, in advance.
- *Boko Haram is not reported to have unspecified political aspirations and objectives.* When Boko Haram is not reported to have unspecified political goals, and other conditions hold true, we can predict the occurrence of abductions one, two and four months in advance using a TP-Rule we were able to derive.
- *The government reportedly shut down Boko Haram's offices.* By examining months where the Nigerian Government shut down Boko Haram's offices, and when other conditions were met, we were able to drive TP-Rules for predicting abductions two and four months in advance.

- *The leadership of Boko Haram was not reportedly installed by an external power.* In months where some other conditions are met and the leadership of Boko Haram was not reportedly installed by an external power, we were able to derive TP-Rules to predict abductions two months in advance.
- *Boko Haram is not reportedly releasing recruitment propaganda.* When Boko Haram is not reported to be releasing propaganda for recruitment purposes, we were able to derive TP-Rules in conjunction with other variables to predict abductions two months ahead of time.
- *Boko Haram did not reportedly provide another NSAG with military aid.* When Boko Haram is not reportedly providing other non-state armed groups (NSAGs)

with military aid, and certain other variables are satisfied, we were able to derive TP-Rules to predict abductions one and two months advance.

- *The government was reportedly not elected, postponed, cancelled elections.* In months where the government was not elected, and other conditions are true, we were able to derive TP-Rules for predicting abductions one and two months ahead of time.
- *Boko Haram did not employ abusive language in its communications or propaganda.* In months where the above variable, and others, hold true, we were able to derive TP-Rules to predict abductions two and four months in advance.
- *Boko Haram's communications and propaganda did not address the public.* When Boko Haram does not address the public and certain other variables are satisfied in the same month, we are able to predict abductions two months in advance using the TP-Rules we derived.
- *The Absence of Explicitly Advocating for Religious Rule.* Another key variable was not explicitly promoting religious rule or a religious government. Though Boko Haram is consistently interested in promoting religious rule and sharia law, their public statements about it are frequent in some months and infrequent and/ or absent in others. Using this variable, in combination with others, we were able to derive several TP-Rules to predict the occurrence of abductions.

All of the previously mentioned variables are those that occur most frequently when predicting abductions two months in advance. The following variable is less frequently used but is also explored in this chapter.

- *Government security forces have not reportedly executed anyone or the practiced had been abolished.* When the government of Nigeria was not reported to have executed anyone, we were able to derive TP-Rules with the help of other variables. These TP-Rules can be used to predict abductions two months in advance.

In addition to all of the variables used to predict the occurrence of abductions, this chapter also examines variables linked to when Boko Haram releases prisoners. Some the of the variables used to derive TP-Rules for predicating prisoner releases are briefly explored below.

- *The Nigerian government did not engage in sexual violence against civilians.* We were able to derive TP-Rules for predicting the release of abducted prisoners when the Nigerian government's forces were not reported to have committed sexual violence and other variables were satisfied. Using these TP-Rules we can predict prisoner releases two and four months in advance.
- *Boko Haram did not issue messages addressing its aspirations and objectives.* Using a TP-Rule derived from when Boko Haram is not communicating about its aspirations and objectives and other conditions are met, we are able to derive TP-rules to predict prisoner releases one and four months in advance.
- *Members of Boko Haram were not reportedly on trial.* When members of Boko Haram are not on trial, and other variables are satisfied, we were able to derive TP-Rule for predicting prisoner releases one and four months in the future.

- *Boko Haram reportedly did not use a pseudonym.* Using months where Boko Haram did not reportedly operate under a pseudonym, and other variables were true, we were able to derive TP-Rules for predicting prisoner releases. These TP-Rules can predict releases two months in advance.

6.1 Abductions and Boko Harm's Use of Child Soldiers and a Second Variable

In months where Boko Haram is reportedly using child soldiers, and another variable is satisfied, we were able to derive a TP-Rule for predicting abductions one month in advance. Some of the variables paired with Boko Haram's use of children include, but are not limited to: no representative(s) of Boko Haram reportedly stood trial, Boko Haram did not issue messages addressing the government, the Nigerian government did not release individual member(s) of Boko Haram, Boko Haram is not using news media and periodicals to get their message out, Boko Haram did not denies or stopped drafting members from certain socio-economic backgrounds, and the government was not reportedly elected, and elections were not postponed or cancelled. In the case of TP-Rule AB-1 (abductions) below, the second variable used examines if Boko Haram was not explicitly reported to be advocating for religious rule.

> **TP-Rule AB-1**
> Abductions occur in months in which:
>
> - 1 month earlier, Boko Haram recruits, trains and/or deploys individuals of young age.
> - 1 month earlier, Boko Haram is not reported as striving for religious rule.
>
>   *Support = 0.19*
>   *Probability = 65%, Inverse Probability = 63%, Negative Probability = 16%, Lift = 1.78*

An example of TP-Rule AB-1 in action can be seen in Febuary of 2015. In that month, Boko Haram was reportedly using child soldiers and the group was not reportely advocating reilgous rule. Then in March 2015, Boko Haram abducted an estimated 500 children from Damasak, Nigeria (Sieff 2017). TP-Rule AB-1 also applies for making predictions two months in advance as well.

6.2 Abduction Predictions When Boko Harm Uses Child Soldiers and Two Other Conditions Are Met

We were able to predict the occurrence of abductions when Boko Haram was reportedly using child soldiers and two other variables were satisfied in the same month. The additional variables include: the Nigerian government did not declare a State of Emergency, a previously installed embargo is reportedly lifted, and Boko Haram did not issue messages addressing its hostages. The third variable is satisfied when: Boko Haram did not use email to communicate, and Boko Haram did not issue messages addressing its campaign. It should be noted that neither of these lists satisfying variables is not exhaustive, and many combinations of variables can be used when Boko Haram is using child soldiers to form TP-Rules. TP-Rule AB-2 is displayed below and looks at when Boko Haram is not reported as striving for unspecified political aspirations and objectives, and the group is not reportedly providing military aid to another NSAG.

TP-Rule AB-2
Abductions occur in months in which:

- 1 month earlier, Boko Haram recruits, trains and/or deploys individuals of young age.
- 1 month earlier, Boko Haram is not reported as striving for unspecified political aspirations and objectives.
- 1 month earlier, Boko Haram did not reportedly give military support to another NSAG.

   Support = 0.19
   Probability = 65%, *Inverse Probability* = 63%, *Negative Probability* = 16%, *Lift* = 1.78

One example of TP-Rule AB-2 can be seen in April 2014 when all three of the conditions required by AB-2 were met. The next month, Boko Haram fighters were suspected to have abducted eight girls from the village of Warabe in Nigeria; the girls were between twelve and fifteen years old (BBC News 2014a).

Another TP-Rule, AB-3, also looks at whether Boko Haram is using child soldiers – however, one of the two supporting variables is different. AB-3 is able to predict abductions one month in advance when Boko Haram is reported to be using child soldiers, security forces have not reportedly executed anyone or the practice has been abolished, and Boko Haram did not reportedly give military support to another NSAG. TP-Rule AB-3 is presented below.

TP-Rule AB-3
Abductions occur in months in which:

- 1 month earlier, Boko Haram recruits, trains and/or deploys individuals of young age.
- 1 month earlier, security forces have not reportedly executed anyone or the practice has been abolished.
- 1 month earlier, Boko Haram did not reportedly give military support to another NSAG.

Support = 0.20
Probability = 67%, Inverse Probability = 67%, Negative Probability = 15%, Lift = 1.78

An example of rule AB-3 occurred in July of 2014. In this month, Boko Haram was reported to be using child soldiers, the group was not reportedly providing military aid to another NSAG and security forces did not reportedly execute anyone. Then in August, Boko Haram attacked a village in northern Cameroon and kidnapped the child of the local chief; ten people were also killed in the terrorists' raid (BBC News 2014b). There are too many other TP-Rules that use Boko Haram's use of child soldiers as a variable to include them all in this chapter. The variables paired with Boko Haram's use of child soldiers may be different, but the offsets and results for the rules are similar.

6.3   Abductions When Government Shuts Down Boko Haram's Offices, the Group Does Not Address the Public, and the Government Was Reportedly Not Elected

Unlike the previous rules, TP-Rule AB-4 does not consider the situation when Boko Haram is using child soldiers. Rule AB-4 is able to predict abductions two months after the government closes Boko Haram's offices, Boko Haram did not issue messages addressing the public, and the government was reportedly not elected, or elections were postponed or cancelled. One example can be seen in July of 2016 when all of the conditions in Rule AB-4 were met. Two months later an abduction by Boko Haram was reported. The results of TP-Rule AB-4 are displayed below.

It should be noted that many TP-Rules look at when the Nigerian government closes some Boko Haram locations. The other variables can differ, but the results and prediction statistics and temporal offsets are similar. From now on, this chapter will transition to examine the circumstances when Boko Haram releases prisoners.

TP-Rule AB-4
Abductions occur in months in which:

- 2 months earlier, the Nigerian government closes some Boko Haram locations.
- 2 months earlier, Boko Haram did not issue messages addressing the public.
- 2 months earlier, the government was reportedly not elected, postponed, cancelled elections.

Support = 0.19
Probability = 65%, Inverse Probability = 63%, Negative Probability = 16%, Lift = 1.76

6.4   Boko Haram's Release of Prisoners

In months where Boko Haram is not reportedly using child soldiers, we are able to make predictions about prisoner releases three months in advance using a TP-Rule we derived. TP-Rule AB-5 below depicts the results of using Boko Haram's deployment of child soldiers to predict prisoner releases.

TP-Rule AB-5
Prisoner releases occur in months in which:

- 3 months earlier, Boko Haram recruits, trains and/or deploys individuals of young age.

Support = 0.21
Probability = 67%, Inverse Probability = 75%, Negative Probability = 10%, Lift = 2.61

An example of AB-5 can be seen in July of 2014 when Boko Haram was reported to be using child soldiers. Three months later in October, Boko Haram released 27 people it had been holding hostage. Among them were ten Chinese nationals and the wife of Cameroon's Vice Prime Minister (BBC News 2014c). It should be noted that rule AB-5 can also be used to predict prisoner releases five months in advance.

6.5   Boko Haram's Release of Prisoners When the Group Uses Child Soldiers and a Second Variable Is True

Like TP-Rule AB-5, TP-Rule AB-6 uses Boko Haram's use of child soldiers to predict prisoner releases but differs by examining a variety of second variables. These second variables include, but are not limited to: Boko Haram did not reportedly have foreign members, no direct negotiations between Boko Haram and a state are currently taking place, Boko Haram did not issue messages addressing claims of responsibility, and the Nigerian government did not declare a state of emergency. The second variable chosen determines how far into the future the TP-Rule is able to predict. In the case of AB-6, the second variable looks at when members of Boko Haram are not reportedly on trial. This variable paired with Boko Haram's use of child soldiers allows the rule to predict one and four months in the future. The depiction of AB-6 below presents results for predicting one month in advance.

TP-Rule AB-6
Prisoner releases occur in months in which:

- 1 month earlier, Boko Haram recruits, trains and/or deploys individuals of young age.
- 1 month earlier, members of Boko Haram were not reportedly on trial.

    Support = 0.21
    Probability   =   67%,   Inverse   Probability   =   75%,   Negative Probability = 10%, *Lift* = 2.61

One example of TP-Rule AB-6 occurred in September of 2016 when Boko Haram militants were not reportedly on trial and the group was reported to be using child soldiers. One month later in October, 21 of the Chibok girls, originally abducted over two years prior in 2014 were released into the custody of the Nigerian government (The Guardian 2017).

6.6   Boko Haram's Release of Prisoners When the Group Is Using Child Soldiers, No Members of the Group Are on Trial, and Boko Haram Is Not Using Pseudonyms

In months where Boko Haram is reportedly using child soldiers, members of the group are not reported to be on trial, and Boko Haram is not operating under pseudonyms, we are able to predict prisoner releases two months in advance using a TP-Rule we derived. This rule is TP-Rule AB-7 and it is displayed below.

TP-Rule AB-7
Prisoner releases occur in months in which:

- 2 months earlier, Boko Haram recruits, trains and/or deploys individuals of young age.
- 2 months earlier, members of Boko Haram were not reportedly on trial.
- 2 months earlier, Boko Haram did not reportedly use a pseudonym.

> *Support* = 0.17
> *Probability* = 65%, *Inverse Probability* = 63%, *Negative Probability* = 14%, *Lift* = 2.64

An example of AB-7 can be seen in May of 2014 when all the above conditions were true. Then in June, Boko Haram released two Italian priests and a Canadian nun into the Government of Cameroon's custody (France 24 2014).

6.7   Boko Haram's Release of Prisoners When the Group Is Using Child Soldiers, the Government Did Not Commit Acts of Sexual Violence, and Boko Haram Did Not Issue Messages About Its Aspirations and Objectives

TP-Rule AB-8 is used to predict prisoner releases four months ahead of time. This rule was derived by examining three variables. The first variable was whether Boko Haram used child soldiers during the course of a month. The second variable examines if the Nigerian government was not reported to have committed acts of sexual violence, and the third checks if Boko Haram did not release communications about its aspirations and objectives. If all of these events happen in the same month then AB-8 can be used. The results of AB-8 are presented below.

TP-Rule AB-8
Prisoner releases occur in months in which:

- 4 months earlier, Boko Haram recruits, trains and/or deploys individuals of young age.
- 4 months earlier, the Nigerian government did not engage in sexual violence against civilians.
- 4 months earlier, Boko Haram did not issue messages addressing its aspirations and objectives.

> *Support* = 0.19
> *Probability* = 67%, *Inverse Probability* = 67%, *Negative Probability* = 13%, *Lift* = 2.58

An example of rule AB-8 occurring can be seen in December 2015 when all of the above variables were satisfied in the same month. Then in April 2016, 275 people were freed during Nigerian military operations in northeastern Nigeria (Musa and Tsokar 2015). It should be mentioned that Boko Haram's use of child soldiers is a component of many TP-Rules predicting prisoner releases. Like the rules regarding abductions, there are too many rules involving child soldiers to include all of them in this chapter.

6.8   Conclusions

Boko Haram's use of abductions has terrorized the people of Northeastern Nigeria for years. Parents have their children taken from them, potentially never to see their child again. The kidnapped children face sexual abuse and are often forced to become suicide bombers. Those kidnapped by the group often remain imprisoned for years until they are released or escape on their own. Women forcibly married to fighters face a stigma of being a member of the group rather than a victim and are often rejected by their communities (Bradford 2017). The results presented in this section offer a method to help prevent these terrible crimes from being committed. They also offer a window into when we can expect Boko Haram to release the people they have kidnapped. The following paragraphs describe some of the conditions we have linked to Boko Haram's abductions. It is important to note these relationships are not causative.

- Boko Haram's use of child soldiers is linked to both abductions and prisoner releases in the following months.
- When Boko Haram is not actively and explicitly advocating for religious rule and the group is using child soldiers, abductions appear to follow one and two months into the future.
- Boko Haram seems to abduct people one month after the group has reportedly been using child soldiers, the group is not reported to be striving for unspecified political aspirations, and Boko Haram did not reportedly give military support to another group.
- When Boko Haram uses child soldiers, government security forces did not execute anyone, and Boko Haram did not give military support to another non-state armed group, abductions can be expected to follow in the next month.
- Abductions can be expected two months after the Nigerian government closes some Boko Haram locations, Boko Haram did not issue messages addressing the public, and Nigerian elections were cancelled, postponed, or not held.

While previously mentioned examples relate to Boko Haram's use of abductions, and the following bullets describe the conditions associated with when Boko Haram releases prisoners. Again, these relationships are not causative.

- Boko Haram tends to release prisoners one month after months in which the group deploys child solders and members of Boko Haram were not reportedly on trial.
- When Boko Haram uses child soldiers, members of the group were not reportedly on trial, and Boko Haram did not reportedly use a pseudonym in the same month, prisoners' releases can be expected two months later.
- When Boko Haram uses child soldiers, Nigerian security forces did not commit sexual assaults, and Boko Haram did not release communications about its objectives, the group can be expected to release prisoners four months later.

All of these relationships are not causative, but they are valuable. They provide a window through which we can gain a better understanding of how Boko Haram operates. The rules we derived allow us to better anticipate the actions of Boko Haram and hopefully prevent the abductions which have helped make the group notorious. If abductions can be prevented, women will be spared the horrific treatment they face at the hands of the group and children can be spared from being turned into suicide bombers. The rules presented in this chapter offer security forces some forewarning in an asymmetric conflict against a ruthless opponent.

6.9   Predictive Model/Reports Results

Figure 6.3 below shows the performance of our predictive model and predictive reports during 2019. From January 2019 to the writing of this book, our team has produced reports containing our predictions on the likelihood of several different attacks out by Boko Haram. Our predictive model uses over 90 months' worth of data collected for our research. The dataset is updated at the end of every month to reflect new occurrences in the Boko Haram conflict. We used 6 classifiers on our dataset: SVM, KNN, Random Forrest, Gaussian Naïve Bayes, Multinomial Naïve Bayes, and Logistic Regression. The model aims to predict whether abductions will, or will not, happen within a given timeframe. For example, if the offset is 2 then the model predicts whether or not an event will occur any time during the next two months. The Fig. 6.3 below depicts the results of our predictions compared to the ground truth observed after the predictions we made.

Abduction						
Time Period	1	2	3	4	5	6
Recall	90%	91%	100%	100%	100%	100%
Precision	90%	100%	100%	100%	100%	100%
Accuracy	82%	91%	100%	100%	100%	100%
F1	0.90	0.95	1.00	1.00	1.00	1.00

Fig. 6.3 Abductions results for our 2019 predictive reports

References

Abubakar A, Levs J (2014) 'I will sell them,' Boko Haram leader says of kidnapped Nigerian girls. CNN. https://www.cnn.com/2014/05/05/world/africa/nigeria-abducted-girls/index.html. Accessed 4 Feb 2019

BBC News (2014a) More Nigerian girls abducted by suspected Boko Haram militants. https://www.bbc.com/news/world-africa-27298614. Accessed 6 Feb 2019

BBC News (2014b) 'Islamist militants' kill 10 in northern Cameroon. https://www.bbc.com/news/world-africa-28684302. Accessed 6 Feb 2019

BBC News (2014c) Boko Haram releases 27 hostages in Cameroon. https://www.bbc.com/news/av/world-africa-29583820/boko-haram-releases-27-hostages-in-cameroon. Accessed 6 Feb 2019

BBC News (2017) Nigeria Chibok abductions: what we know. https://www.bbc.com/news/world-africa-32299943. Accessed 4 Feb 2019

BBC News (2018) Nigeria Dapchi abductions: schoolgirls finally home. https://www.bbc.com/news/world-africa-43535872. Accessed 4 Feb 2019

Bradford A (2017) Former Boko Haram captives face stigma, often from other survivors. Huffington Post. https://www.huffingtonpost.com/entry/boko-haram-survivors-face-stigma_us_58e7f434e4b058f0a02f3c0f. Accessed 7 Feb 2019

France 24 (2014) Italian priests, Canadian nun freed in Cameroon. https://www.france24.com/en/20140601-kidnapped-italian-priests-canadian-nun-freed-cameroon. Accessed 7 Feb 2019

Freeman C (2018) Ransom money has turned Boko Haram into Nigeria's Cosa Nostra. The Spectator. https://blogs.spectator.co.uk/2018/04/ransom-money-has-turned-boko-haram-into-nigerias-cosa-nostra/. Accessed 4 Feb 2019

Musa N, Tsokar K (2015) Military kills 15 terrorists, rescues 275 hostages in Borno. The Guardian. https://guardian.ng/news/military-kills-15-terrorists-rescues-275-hostages-in-borno/. Accessed 7 Feb 2019

PBS News Hour (2016) What happened to 10,000 boys kidnapped by Boko Haram? https://www.pbs.org/newshour/show/happened-10000-boys-kidnapped-boko-haram. Accessed 4 Feb 2019

Sieff K (2017) The growing horror in a city where 500 children were kidnapped by Boko Haram. The Washington Post. https://www.washingtonpost.com/world/africa/the-growing-horror-in-a-city-where-500-children-were-kidnapped-by-boko-haram/2017/04/26/59df0916-29e5-11e7-9081-f5405f56d3e4_story.html. Accessed 4 Feb 2019

The Guardian (2017) Boko Haram releases dozens of Chibok schoolgirls, say Nigerian officials. https://www.theguardian.com/world/2017/may/06/boko-haram-releases-dozens-of-kidnapped-chibok-schoolgirls. Accessed 4 Feb 2019

UNICEF (2018) More than 1,000 children in northeastern Nigeria abducted by Boko Haram since 2013. https://www.unicef.org/wca/press-releases/more-1000-children-northeastern-nigeria-abducted-boko-haram-2013. Accessed 4 Feb 2019

Vanguard (2018) How UNIMAD lectures, 10 policewomen were released by Boko Haram. https://www.vanguardngr.com/2018/02/unimad-lectures-10-policewomen-released-boko-haram/. Accessed 4 Feb 2019

Chapter 7
Arson

Boko Haram has used arson as one of its vehicles of terror since the group turned to violence in 2009. From 2011 onward, the group has steadily increased its use of arson in its campaign of terror. Boko Haram's use of arson peaked in 2014 when an arson attack occurred in every month of the year. During 2015 and 2016 Boko Haram's use of arson fell off slightly with 9 months of each year containing an arson attack. Figure 7.1 depicts Boko Haram's use of arson attacks per year from 2009 to 2016.

The arson attacks committed by Boko Haram have devastating effects on those affected by them. Many people are killed in such attacks and those who survive are left homeless. Survivors driven from their homes often have no better option than to go to one of the internally displaced people (IDP) camps set up by the Nigerian government. People often do not have enough to eat in these camps and their living quarters are far from adequate. On top of these issues, children often have limited access to an education (60 Minutes Overtime 2019). Some instances of Boko Haram's use of arson are listed below.

- **February – March 2012**. During February and March, Boko Haram razed at least nine schools in several arson attacks (Human Rights Watch 2012).
- **May 21, 2014**. Boko Haram attacked the village of Shawa and killed ten people. During the attack, the group set fire to the entire village (BBC News 2014b).
- **June 4, 2014**. In an attack on a church in Attagara Nigeria, twenty people were killed, and buildings were burned down by Boko Haram fighters (BBC News 2014a).
- **January 3, 2015**. The village of Baga and surrounding areas were attacked by Boko Haram. They sustained damage from arson attacks (Segun 2015).
- **December 25, 2015**. In a Christmas Day attack, Boko Haram burnt the village of Kimba to the ground and killed fourteen people (France-Presse 2015).

© The Author(s), under exclusive license to Springer Nature Switzerland AG 2021
V. Subrahmanian et al., *A Machine Learning Based Model of Boko Haram*,
Terrorism, Security, and Computation, https://doi.org/10.1007/978-3-030-60614-5_7

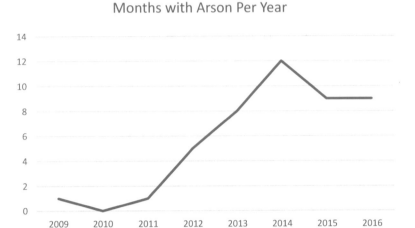

Fig. 7.1 Number of Months Per Year when Boko Haram Carried out Acts of Arson (2009–2016)

- **February 1, 2016**. The village of Dalori was attacked by Boko Haram and 80 people were killed as a result. Many people were shot or burned alive in their homes (The Guardian 2016).
- **August 25, 2017**. Eleven people were killed and thirty homes were torched after Boko Haram fighters attacked Gakara, a village in Cameroon (Aljazeera 2017).
- **November 1, 2018**. Members of Boko Haram attacked the villages of Bulaburin and Kofa as well as an internally displaced persons (IDP) camp in Dalori. The group burnt both villages to the ground and half of the IDP camp was destroyed. Twelve people died in the attack (Aljazeera 2018).
- **November 15, 2018**. Boko Haram attacked Mamanti village in rural northeastern Nigeria. Several homes were burned and two people were wounded in the attack (Murray and Nossiter 2011).

Boko Haram's arson attacks have destroyed and displaced many of Nigeria's northeastern communities. People's lives, livelihoods, and homes have all been lost in the group's campaign of terror. This chapter's focus is on TP-Rules we derived to predict when Boko Haram will, or will not, commit arson attacks. We will now briefly discuss the factors that played a significant role in the TP-rules that we derived. The variables most used for developing TP-Rules which predict an absence of arson attacks one month in advance can be seen in Fig. 7.2. These variables, as well as others used in this chapter, are briefly explained below.

- *The government did not reportedly shut down Boko Haram's offices.* By examining months where the Nigerian Government did not shut down Boko Haram's offices, and when other conditions were met, we were able to derive TP-Rules for predicting abductions two and four months in advance.
- *Boko Haram was not reportedly recruiting, training and/or deploying individuals of young age.* When there are no reports about Boko Haram's use of child

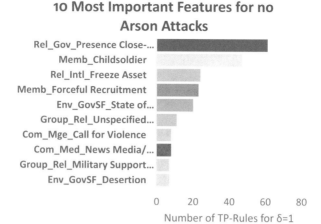

Fig. 7.2 Most Important Variables Linked to Arson Attacks

soldiers, and other variables are satisfied, we are able to derive several TP-Rules. Using these rules, we can reliably predict a lack of arson attacks one, two, three, four, and five months in advance.

- *International organizations did not reportedly freeze Boko Haram's assets.* In months where Boko Haram's assets were not reportedly frozen by a foreign organization, and other variables were satisfied, we were able to derive TP-Rules to predict an absence of arson attacks. Using these rules, we can predict an absence of arson attacks, one, two, and three months in advance.

- *Boko Haram is not reportedly using force to recruit new members.* When Boko Haram is not reportedly using force to recruit new members, and other conditions are met, we are able to derive TP-Rules to predict a lack of arson attacks one, two, three, and four months in advance.

- *Government security forces have not reportedly declared a state of emergency.* In months where a state of emergency has not been declared, and other variables are satisfied, we were able to derive several TP-Rules to predict a lack of arson attacks. These rules allow us to predict an absence of arson attacks one, three, and five months in advance.

- *Boko Haram did not reportedly provide unspecified aid to another NSAG.* In months where other variables were satisfied, and Boko Haram did not reportedly provide any other non-state armed group with unspecified aid, we were able to derive several TP-Rules to predict lack of arson attacks. The rules allow us to predict a lack of arson attacks one month ahead of time.

- *Boko Haram did not reportedly release communications or propaganda calling for violence.* In months where Boko Haram did not reportedly release messages calling for violence, and other variables are satisfied, we were able to derive TP-Rules to predict a lack of arson attacks. These rules allow us to predict a lack of arson attacks one month in advance.

- *Boko Haram did not reportedly use the news mediums or periodicals to release communications or propaganda.* When Boko Haram did not reportedly use news sources or periodicals to spread its propaganda, and certain other conditions were met, we were able to derive several TP-Rules to predict an absence of arson attacks one month in advance.
- *Boko Haram did not provide military support to another NSAG.* For a given month, if Boko Haram did not reportedly provide another NSAG with military aid, and other conditions were met, we were able to derive TP-Rules to predict an absence of arson attacks. These TP-Rules can be used to predict a lack of arson attacks one and two months in advance.
- *Government security forces did not reportedly desert.* When there were no reports of the government's security forces deserting and certain other variables are satisfied, we were able to derive TP-Rules that predict a lack of arson attacks one and two months in advance.

The above features do not constitute all of the features used to derive all of the TP-Rules. Some of the other features we used to predict a lack of arson attacks are briefly explained below.

- *Boko Haram was not reportedly designated as an international terrorist organization.* In months in which Boko Haram is not designated as an international terrorist organization, and certain other conditions are met, we were able to derive TP-Rules to predict a lack of arson attacks. These rules can be used to predict a lack of arson attacks one month in advance.
- *International organizations did not accuse Boko Haram of human rights violations.* We were able to derive TP-Rules predicting a lack of arson attacks from months where international organizations did not accuse Boko Haram of human rights abuses. These TP-Rules also depended on other variables being satisfied, and the rules could be used to predict a lack of arson one and three months in advance.
- *Government Security Forces did not reportedly commit acts of sexual violence.* In months when security force personnel do not reportedly commit acts of sexual violence, and certain other conditions are met, we are able to derive TP-Rules to predict a lack of arson attacks. These rules allow us to make predictions five months in advance.

In addition to the above features used for predicting an absence of arson attacks, below are a couple of the features we used to derive a rule to predict the occurrence of arson attacks.

- *The Absence of Explicitly Advocating for Religious Rule.* Another key variable was not explicitly promoting religious rule or a religious government. Though Boko Haram is consistently interested in promoting religious rule and sharia law, their public statements about it are frequent in some months and infrequent and/or absent in others. Using this variable, in combination with others, we were able to derive a few TP-Rules to predict the occurrence of arson attacks one month in advance.

- *Government security forces did not execute captured Boko Haram terrorists.* In months where government security forces did not reportedly execute Boko Haram terrorists, and Boko Haram was not actively promoting religious rule, we were able to derive a TP-Rule to predict arson attacks one month in advance.

7.1 Arson Attacks When Boko Haram Is Not Explicitly Striving for Religious Rule and a Second Variable Is True

Unlike the other chapters, this chapter will only discuss one rule about predicting when an event will occur. This section discuses TP-Rule AR-1 (Arson), a rule which can be used to predict arson attacks one month ahead of time. This rule requires Boko Haram to not be actively promoting religious rule, and for one other variable to be satisfied in the same month. These other variables include: Boko Haram not supporting other NSAGs financially, Boko Haram is not engaged in nonviolent conflict with any other NSAG, and Boko Haram has not clearly stated who its enemies are. In AR-1 below, the second variable examines when government security forces are not reportedly executing members of Boko Haram.

> **TP-Rule AR-1**
> Arson attacks occur in months in which:
>
> - 1 month earlier, Boko Haram is not explicitly striving for religious rule.
> - 1 month earlier, government security forces are reportedly not executing captured Boko Haram militants.
>
>   *Support = 0.39*
>   *Probability = 67%, Inverse Probability = 80%, Negative Probability = 23%, Lift = 1.78*

An example of AR-1 can be seen in January of 2016. During that January, both of the above conditions were met, and one-month later Boko Haram used arson in one of its attacks. Boko Haram attacked the village of Dalori in February 2016. In this raid, the group killed 86 people after bombing the village and burning it to the ground (The Guardian 2016). The remaining rules discussed in this chapter examine TP-Rules which predict an absence of arson attacks.

7.2 A Lack of Arson Attacks, Boko Haram's Use of Child Soldiers and Terrorist Organization Designation

Section 7.2 displays a TP-Rule that is used to predict an absence of arson attacks. Rule AR-2 requires Boko Haram to not reportedly be using child soldiers, and for the group not be designated as an international terrorist organization in the same month. If both of these conditions are satisfied, TP-Rule AR-2 can be used to predict a lack of arson attacks one month in advance. The rule can be viewed below.

TP-Rule AR-2
Arson attacks do not occur in months in which:

- 1 month earlier, Boko Haram was not reportedly employing child soldiers.
- 1 month earlier, Boko Haram was not designated as an international terrorist organization.

   Support = 0.44
   Probability = 75%, Inverse Probability = 86%, Negative Probability = 16%, Lift = 1.78

An example of rule AR-2 can be seen in September of 2011 when both of the above conditions were satisifed. Then in that October, no arson attacks were committed by Boko Haram. It should also be noted that rule AR-2 can be used to make predictions, two, three, four, and five months in advance.

7.3 A Lack of Arson Attacks, the Closure of Boko Haram's Sites and the Group's Terrorist Designation

The rule displayed in this section looks at when the government does not reportedly close down sites operated by Boko Haram and when the group was not designated as an international terrorist organization. In a month where both of these conditions are met, TP-Rule AR-3 can be used to predict a lack of arson attacks one month in advance. This rule can be seen below.

One example of this rule can be seen in April of 2010. In this month, both of the conditions were true, and no arson attacks were reported in May 2010. In addition, rule AR-3 can be used to predict a lack of arson attacks two, three, and five months in advance. Using the April of 2010 to satisfy this TP-Rule, rule AR-3 was also able to predict the absence of arson attacks in June, July, and September of that year.

TP-Rule AR-3
Arson attacks do not occur in months in which:

- 1 month earlier, the government did not reportedly shut down Boko Haram's offices.
- 1 month earlier, Boko Haram was not designated as an international terrorist organization.

Support = 0.42
Probability = 70%, Inverse Probability = 82%, Negative Probability = 22%, Lift = 1.78

7.4   A Lack of Arson Attacks, Boko Haram's Use of Child Soldiers and a Second Variable

Like rule AR-2, TP-Rule AR-4 looks at when Boko Haram is reportedly not using child soldiers. Rule AR-4 pairs Boko Haram's uses of child soldiers with one of several other variables. These other variables include but are not limited to: the government did not reportedly shut down Boko Haram's offices, no direct negotiations between Boko Haram and the Government of Nigeria were taking place, Boko Haram's communications did not address the public, International organizations reportedly did not seek, deny or revoke prior allegations of war crimes against Boko Haram, and International organizations reportedly did not, deny or release prior frozen assets of members of Boko Haram. Rule AR-4 is displayed below and looks at whether or not the Nigerian government declared a state of emergency.

TP-Rule AR-4
Arson attacks do not occur in months in which:

- 1 month earlier, Boko Haram was not reportedly employing child soldiers.
- 1 month earlier, the Nigerian government did not declare a State of Emergency.

Support = 0.38
Probability = 83%, Inverse Probability = 76%, Negative Probability = 23%, Lift = 1.78

The rules that can be generated by these variables can also be used to predict a lack of arson attacks two months in advance. An example of rule AR-4 in the real world can be seen in 2012. In that November, Boko Haram was not reportedly using child soldiers and no state of emergency was declared. In December of that year no arson attacks were reported. In addition, that January also saw no arson attacks demonstrating the rule's viability for making predictions two months ahead of time.

7.5 A Lack of Arson Attacks, Boko Haram's Use of Child Soldiers and Accusations of Human Rights Violations

Section 7.5 examines TP-Rule AR-5, a rule which looks at Boko Haram's use of child soldiers, and when international organizations have accused the group of human rights violations. When Boko Haram is not reportedly using child soldiers, and international organizations have not accused the group of human rights violations in the same month, AR-5 can be used to predict a lack of arson attacks one and three months in advance. TP-Rule AR-5 with a one-month offset can be seen below.

> **TP-Rule AR-5**
> Arson attacks do not occur in months in which:
>
> - 1 month earlier, Boko Haram was not reportedly employing child soldiers.
> - 1 month earlier, International organization(s) reportedly did not seek, deny or revoke prior allegations of human rights violations against Boko Haram.
>
>   *Support* = 0.42
>   *Probability* = 70%, *Inverse Probability* = 82%, *Negative Probability* = 22%, Lift = 1.78

In July of 2009, Boko Haram was not reported to be using child soldiers, and international organizations had not accused the group of human rights abuses. During that August, no arsons were attributed to the terrorist group.

7.6 A Lack of Arson Attacks When the Government Does Not Shut Down Boko Haram's Offices and Security Forces Do Not Commit Sexual Violence

We were able to derive a TP-Rule that looks at when the government does not shut down Boko Haram's offices and security forces do not commit acts of sexual violence. If both of these variables are satisfied, we can use TP-Rule AR-6 to predict a lack of arson five months in the future. An example of this rule can be seen in April of 2013 when both of the above conditions were met. Then in September there were no reported arson attacks, even though arson attacks were reported in August and October. Rule AR-6 is displayed below.

TP-Rule AR-6
Arson attacks do not occur in months in which:

- 5 months earlier, the government did not reportedly shut down Boko Haram's offices.
- 5 months earlier, there were no reports of sexual violence by Nigerian security forces.

 Support = 0.39
 Probability = 66%, Inverse Probability = 80%, Negative Probability = 23%, Lift = 1.70

7.7   A Lack of Arson Attacks When the Government Does Not Shut Down Boko Haram's Offices and a Second Variable

Like some of the other rules presented in this chapter, TP-Rule AR-7 looks at months where the government has not shut down Boko Haram's offices. The rule requires for one of several other variables to be true. These variables include but are not limited to: International organizations did not reportedly seek, deny or revoke prior allegations of human rights violations against Boko Haram, and Boko Haram did not reportedly, deny or stopped to forcefully recruit people. In the case of AR-7 below, the rule examines months where international organizations did not reportedly freeze assets of Boko Haram.

TP-Rule AR-7
Arson attacks do not occur in months in which:

- 3 months earlier, the government did not reportedly shut down Boko Haram's offices.
- 3 months earlier, International organizations did not reportedly freeze assets of members of Boko Haram.

 Support = 0.40
 Probability = 66%, Inverse Probability = 79%, Negative Probability = 24%, Lift = 1.74

An example of AR-7 can be viewed in June of 2012 since no arson attacks were reported in that month. Three months prior in March, there were no reports of Boko Haram's assets being frozen, nor were closedowns conducted by the government.

7.8 A Lack of Arson Attacks When the Government Is Receiving Not Military Aid and a Second Variable

The final TP-Rule that will be examined in this chapter looks at months where the Nigerian Government is not receiving military aid. In addition, one of several variables must be satisfied for this rule, AR-8, to be valid. These other variables include but are not limited to: Boko Haram reportedly did not, stopped or denied employing child soldiers, and the government did not reportedly shut down Boko Haram's offices. For AR-8 displayed below, the second variable looked at when the Nigerian government did not declare a State of Emergency.

> **TP-Rule AR-8**
> Arson attacks do not occur in months in which:
>
> - 2 months earlier, military aid to the government has been suspended or denied.
> - 2 months earlier, the Nigerian government did not declare a State of Emergency.
>
>   *Support* = 0.41
>   *Probability* = 72%, *Inverse Probability* = 82%, *Negative Probability* = 21%, *Lift* = 1.76

One example of this rule can be seen in January of 2016 when both of the conditions of rule AR-8 were met. Then in March no arson attacks were reported even though nine months of 2016 had arson attacks occur.

7.9 Conclusions

Arson attacks have had devastating consequences for the people of northeastern Nigeria. Many people have been killed by the violence and many more have been displaced for years since their homes were destroyed. Thousands have no better option but to languish in the harsh conditions of the government's Internally Displaced Person camps (Mohamed 2018). Once in these camps, many find it difficult to return home since Boko Haram still poses a threat, or because they do not have the resources to rebuild their lives. The ability to accurately predict arson attacks in advance would enable law enforcement to better deploy resources in order to protect the threatened communities and save peoples' livelihoods. In the following bullets we present some conditions we have associated with Boko Haram's use, or lack of use, of arson. It should be noted that these relationships are not causative.

- When Boko Haram is not explicitly striving for religious rule, and security forces are not reportedly executing Boko Haram militants, arson attacks tend to follow a month later.
- When the group is not using child soldiers, and Boko Haram has not been designated an international terrorist organization, arson attacks do not tend to occur during the following month.
- If the government does not shut down sites run by Boko Haram and the group has not been designated an international terrorist organization, arson attacks seem not to happen one month later.
- When Boko Haram is not using child soldiers and a state of emergency has not been declared, arson attacks tend not to occur one month later.
- Arson attacks do not seem to occur in months after ones where Boko Haram is not reportedly using child soldiers, and the group has not been accused of human rights abuses.
- Five months after there have been no reported acts of sexual violence by security forces, and Boko Haram's sites have not been shut down, arson attacks seem not to occur.
- When assets of Boko Haram have not been frozen, and sites run by the group have not been shut down, arson attacks are absent three months later.
- When the government is not receiving military aid and has not declared a state of emergency, arson attacks seem not to occur two months later.

While these relationships are not causative, they can aid in the fight against Boko Haram. These insights allow us to see when arson attacks are the least likely to occur, and better allocate recourses to combat different actions of Boko Haram. Manpower could be diverted to guard against suicide bombings or prevent abductions. Better yet, when the chance of arson is low, resources could be used to help displaced people rebuild their lives and return home. Defeating Boko Haram and rebuilding parts of northeastern Nigeria is no small task, but any tool to better understand the actions of Boko Haram has significant value.

7.10 Predictive Model/Reports Results

Figure 7.3 shows the performance of our predictive model and predictive reports during 2019. From January 2019 to the writing of this book, our team has produced reports containing our predictions on the likelihood of several different attacks out

Arson						
Time Period	1	2	3	4	5	6
Recall	91%	91%	100%	100%	100%	100%
Precision	100%	100%	100%	100%	100%	100%
Accuracy	91%	91%	100%	100%	100%	100%
F1	0.95	0.95	1.00	1.00	1.00	1.00

Fig. 7.3 Arson results for our 2019 predictive reports

by Boko Haram. Our predictive model uses over 90 months' worth of data collected for our research. The dataset is updated at the end of every month to reflect new occurrences in the Boko Haram conflict. We used 6 classifiers on our dataset: SVM, KNN, Random Forrest, Gaussian Naïve Bayes, Multinomial Naïve Bayes, and Logistic Regression. The model aims to predict whether arson will, or will not, happen within a given timeframe. For example, if the offset is 2 then the model predicts whether or not an event will occur anytime during the next two months. The Fig. 7.3 depicts the results of our predictions compared to the ground truth observed after the predictions we made.

References

60 Minutes Overtime (2019) Beyond the Chibok girls inside Nigeria's Id camps. CBS News. https://www.cbsnews.com/news/beyond-the-chibok-girls-inside-nigerias-idp-camps-60-minutes/. Accessed 2 Mar 2019

Aljazeera (2017) Suspected Boko Haram fighters kill 11 Cameroon. https://www.aljazeera.com/news/2017/08/suspected-boko-haram-fighters-kill-11-cameroon-170825145004120.html. Accessed 2 Mar 2019

Aljazeera (2018) Nigeria: 'Villages totally burned' in deadly Boko Haram attacks. https://www.aljazeera.com/news/2018/11/nigeria-villages-totally-burned-deadly-boko-haram-attacks-181101103918136.html. Accessed 2 Mar 2019

BBC News (2014a) Nigerian villagers 'killed in Boko Haram church attack'. https://www.bbc.com/news/world-africa-27690687. Accessed 2 Mar 2019

BBC News (2014b) Nigeria violence: 'Boko Haram' kill 27 in village attacks. https://www.bbc.com/news/world-africa-27498598. Accessed 2 Mar 2019

France-Presse A (2015) Boko Haram kill at least 14 in Christmas Day attack in Nigeria. https://www.theguardian.com/world/2015/dec/26/boko-haram-kills-at-least-14-in-christmas-day-attack-in-nigeria. Accessed 2 Mar 2019

Human Rights Watch (2012) Nigeria: Boko Haram targeting schools. https://www.hrw.org/news/2012/03/07/nigeria-boko-haram-targeting-schools. Accessed 2 Mar 2019

Mohamed H (2018) Thousands displaced by Boko Haram languish in Abuja IDP camps. Aljazeera. https://www.aljazeera.com/indepth/features/thousands-displaced-boko-haram-languish-abuja-idp-camps-180325070133915.html. Accessed 12 Feb 2019

Segun M (2015) Dispatches: what really happened in Baga, Nigeria? Human Rights Watch. https://www.hrw.org/news/2015/01/14/dispatches-what-really-happened-baga-nigeria. Accessed 2 Mar 2019

The Eagle Online (2018) Boko Haram insurgents resort to arson in North-East. https://theeagleonline.com.ng/boko-haram-insurgents-resort-to-arson-in-north-east/. Accessed 2 Mar 2019

The Guardian (2016) Boko Haram: children among villagers burned to death in Nigeria attack. https://www.theguardian.com/world/2016/feb/01/boko-haram-attack-children-among-villagers-burned-to-death-in-nigeria. Accessed 12 Feb 2019

Chapter 8
Other Types of Attacks

Boko Haram's toolkit of terror tactics includes several attacks that we have not discussed thus far. These attacks include the targeting of government officials, looting,
the targeting of security installations, as well as attempted bombings. These rules
shed light on several attacks used effectively by the group. Boko Haram uses looting
to inflict terror as well as obtain supplies, ammunition, and equipment. Figure 8.1
above shows the frequency of these attacks. This graph shows how the targeting of
security installations was very frequent during the 2012 to 2015 timeframe.
Figure 8.1 also shows how often Boko Haram targeted government officials, carried
out lootings, and engaged in attempted bombings. Attacking security installations
allows the group to demonstrate their operational capacity, inflict casualties, as well
steal military equipment. In addition, the group makes efforts to attack and kill government officials. Finally, not all of Boko Haram's bombings are successful and this
chapter will look at those attacks as well. Some examples of these events include:

- **June 4, 2014**. An attack on a military base in Damboa, Nigeria was repelled by
  government forces (NBC News 2014).
- **November 14, 2014**. Boko Haram militants robbed a bank during a period of
  frequent raids and looting (News24 2014).
- **January 5, 2015**. In a large raid, the group seized a military base in the village
  of Baga (Mark 2015).
- **Jun 18, 2015**. Boko Haram terrorists looted and burned villages in Niger (Vice
  News 2015).
- **November 22, 2016**. Looting took place during the raids of villages near the
  town of Chibok (CBS News 2016).
- **August 25, 2017**. Boko Haram looted and burned villages, stealing livestock and
  grain (Sahara Reporters 2017).
- **November 23, 2018**. Boko Haram attacked a military base and 100 soldiers were
  killed (Kazeem 2018).

© The Author(s), under exclusive license to Springer Nature Switzerland AG 2021   107
V. Subrahmanian et al., *A Machine Learning Based Model of Boko Haram*,
Terrorism, Security, and Computation, https://doi.org/10.1007/978-3-030-60614-5_8

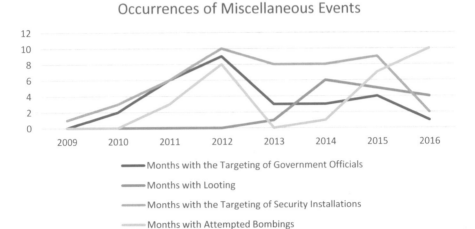

Fig. 8.1 Number of months in which different types of attacks were carried out by Boko Haram (2009–2016)

- **December 28, 2018**. Two military bases were raided by Boko Haram. At least one officer was killed, and the group seized various small arms and vehicles (Al Jazeera 2018).

These diverse types of attacks launched by Boko Haram have played a devastating role in the group's campaign of terror. Every successful looting helps the group resupply food and/or war material, while successful raids on military bases erode the morale of civilians and soldiers alike. This chapter focuses on the TP-Rules that predict when these attacks occur. In the following section, we will discuss the variables involved in these TP-Rules.

- *The government did not reportedly shut down Boko Haram's offices.* One TP-Rule we derived examined months where Boko Haram's operational locations were not shut down by the Nigerian Government. If this condition, along with some others, is met, then the targeting of government officials can be predicted one month ahead of time.
- *Government security forces are reportedly not executing captured Boko Haram militants.* In months where Nigerian Security forces have not reportedly executed anyone, and certain other variables are satisfied, we were able to derive a TP-Rule to predict the targeting of government officials one month out.
- *Nigerian Security forces were not engaged in forced resettlements.* We were able to derive a TP-Rule after examining months were security forces did not forcibly resettle people. This rule could be used to predict the targeting of government officials one month in advance.
- *Military aid to the government has been suspended or denied.* One TP-Rule we derived is able to predict looting three months ahead of time. The rule examines when military aid to the Nigerian government has been suspended or denied.

Another TP-Rule derived using this variable is able to predict the absence of targeting of security installations four months in advance.

- *Boko Haram reportedly did not, denies or stopped utilizing (unspecified) media to release its propaganda.* By examining months where Boko Haram is not using unspecified media to release its propaganda, we were able to derive a TP-Rule to predict looting three months in advance. This rule was derived in conjunction with various other variables as well.
- *Boko Haram did not issue messages with threats.* By examining months where Boko Haram did not issue messages containing threats, we were able to derive a TP-Rule to predict the absence of targeting of security installations four months in advance.
- *Boko Haram was not reportedly supporting other NSAGs financially.* When Boko Haram was not reportedly supporting another NSAG financially during a given month, we are able to use a TP-Rule we derived to predict the absence of targeting of security installations four months in advance. This TP-Rule also required that other variables be satisfied in order to make a valid prediction.
- *Boko Haram did not issue messages with claims of responsibility.* A TP-Rule that we derived is able to predict the absence of targeting of security installations four months in advance. One additional variable that needs to be satisfied for this rule requires that Boko Haram did not issue messages claiming responsibility during a given month.
- *Boko Haram communications did not address the public.* One variable that contributed to a TP-Rule examined when Boko Haram's communications addressed the public. When the group's communications did not address the public, and other variables are satisfied, our TP-Rule can be used to predict the absence of targeting of security installations four months in advance.
- *Boko Haram recruits, trains and/or deploys individuals of young age.* By examining months where Boko Haram uses child soldiers, we were able to derive a TP-Rule to predict attempted bombings. This rule can be used to predict the occurrence of attempted bombings three months in advance.

8.1 The Targeting of Government Officials and Executions by Security Forces, the Closure of Boko Haram's Locations, and a Third Variable

This chapter will focus on a few rules that can be used to predict several different attacks and events. The first of these rules is TP-Rule MS-1 (Miscellaneous), and this rule can be used to predict when government officials will be targeted by Boko Haram. For this rule to be viable, the government must not have shut down locations controlled by the group, government security forces must not execute any captured terrorists, and a third variable must be satisfied. For rule MS-1, the third variable requires that Nigerian Security forces not be involved in forcible resettlement activities. When these three variables are satisfied, MS-1 can be used to predict the targeting of government officials one month in advance. Rule MS-1 can be viewed below.

TP-Rule MS-1
The targeting of government officials occurs in months in which:

- 1 month earlier, the government did not close down Boko Haram's locations.
- 1 month earlier, government security forces are reportedly not executing captured Boko Haram militants.
- 1 month earlier, Nigerian security forces were not engaged in forcible resettlement activities.

  Support = 0.19
  Probability = 65%, Inverse Probability = 61%, Negative Probability = 17%, Lift = 1.78

An example of MS-1 in real life can be seen in September of 2010. During this September, all three conditions required by Rule MS-1 were met. Then during the next month, a government official was targeted during an attack by Boko Haram.

8.2   Looting, the Suspension of Military Aid, the Groups Use of Unspecified Media, and a Third Variable

Another attack frequently used by Boko Haram is looting. Not only does this type of attack damage communities and the capabilities of security forces, it actively replenishes the group's stores of food, ammunition, and weapons. TP-Rule MS-2 can be used to predict the occurrence of lootings three months in advance, and it depends on three variables being satisfied during the same month. For two of these variables to be satisfied, military aid to the government must be suspended or denied, and Boko Haram must not be using unspecified media to disseminate its propaganda. The third variable can be satisfied by any one of several conditions. These conditions include but are not limited to: Boko Haram notably did not employ abusive language or call for recruits in its communication, Boko Haram is not supporting other non-state armed groups (NSAGs for short) military, and Foreign aid to the government has been suspended or denied. For Rule MS-2, the third rule examines Nigerian border closures.

TP-Rule MS-2
Looting occurs in months in which:

- 3 months earlier, military aid to the government has been suspended or denied.
- 3 months earlier, the group was not reportedly utilizing (unspecified) media.
- 3 months earlier, the border has reportedly not been closed or re-opened after closure.

  Support = 0.11
  Probability = 67%, *Inverse Probability* = 67%, *Negative Probability* = 7%, *Lift* = 3.48

An example of this rule can be seen during February of 2014. During this month, all of the variables in rule MS-2 were satisfied and a looting incident occurred in May 2014. One attack during this month saw Boko Haram attack the village of Alagarno in northeastern Nigeria; 17 people were killed during the raid and subsequent looting of the village (BBC News 2014).

8.3 The Absence of Targeting of Security Installations, the Suspension of Military Aid, Threatening Messages, and a Third Variable

Boko Haram has frequently targeted military bases and security installations during its campaign of terror. By targeting security installations, the group is able to both cause security force casualties and loot military hardware after successful engagements. TP-Rule MS-3 was derived to allow us to predict the absence of targeting of security installations four months in advance. For this rule to be viable, three variables must be satisfied. One variable requires that military aid to the Nigerian government be suspended or denied. The second variable is satisfied if Boko Haram does not issue messages containing threats. Rule MS-3's third variable can be satisfied by the occurrence of any one of several conditions and these conditions include, but are not limited to: Boko Haram communications did not address security forces, there was no reported sexual violence by Nigerian security forces, and members of the group did not surrender. For MS-3 below, the third variable is satisfied if Boko Haram was not reportedly supporting other NSAG financially.

An example of this rule can be seen in March of 2011 when no security installations were targeted. Four months prior, all of the conditions in rule MS-3 had been met.

TP-Rule MS-3
Security Installations are Not Targeted in months in which:

- 4 months earlier, military aid to the government had been suspended or denied.
- 4 months earlier, Boko Haram did not issue messages with threats.
- 4 months earlier, Boko Haram was not supporting other NSAGs financially.

   Support = 0.36
   Probability   =   66%,   Inverse   Probability   =   78%,   Negative
   Probability = 23%, Lift = 1.72

8.4 The Absence of Targeting of Security Installations, the Suspension of Military Aid, Claims of Responsibility, and a Third Variable

This section of the chapter also examines a rule that can be used to predict the absence of targeting of security installations by Boko Haram. Like rule MS-3, rule MS-4 requires that military aid to the Nigerian government be either suspended or denied. The second component of rule MS-4 is satisfied if Boko Haram does not issue messages containing a claim of responsibility. The third variable in rule MS-4 can be satisfied by several conditions which include, but are not limited to: there being no reported sexual violence by Nigerian security forces, Boko Haram is not supporting other NSAGs financially, and Boko Haram is not using news media and periodicals to spread their propaganda. Rule MS-4, as shown below, is satisfied when the group's communications do not address the public.

TP-Rule MS-4
Security Installations are Not Targeted in months in which:

- 4 months earlier, military aid to the government had been suspended or denied.
- 4 months earlier, Boko Haram did not issue messages with claim of responsibility.
- 4 months earlier, Boko Haram communications did not address the public.

   Support = 0.36
   Probability   =   65%,   Inverse   Probability   =   78%,   Negative
   Probability = 24%, Lift = 1.72

During February of 2010, all of the conditions required by MS-4 were met and in June of the same year, no security installations were targeted.

8.5 Attempted Bombing and the Use of Child Soldiers

The final rule to be discussed in this chapter examines attempted bombings. Boko Haram frequently uses various bombing techniques in their war on terror. However, not all of the bombings are successful. Suicide bombers have been foiled before they reach their targets, and some bombs fail to detonate or are disarmed. TP-Rule MS-5 can be used to predict the occurrence of these attempted bombings. This rule can be used to make predictions three months in advance if Boko Haram is reportedly using, deploying, or training child soldiers during a given month. Rule MS-5 can be viewed below.

TP-Rule MS-5
Attempted Bombings occur in months in which:

- 3 months earlier, the organization recruits, trains and/or deploys individuals of young age.

  Support = 0.21
  Probability = 67%, Inverse Probability = 62%, Negative Probability = 18%, Lift = 1.74

In September 2015, Boko Haram was reportedly using child soldiers. Then, three months later in December, the group launched a large attack on Maiduguri. During this attack the Nigerian security forces were able to stop 13 suicide bombers before they were able to reach their targets (The Washington Post 2016).

8.6 Conclusions

All of these various attacks committed by Boko Haram contribute to the group's reign of terror. The group's targeting of government officials allows the group to degrade the rule of law and disrupt the communities those leaders served. Looting is a particularly harmful attack perpetrated by the group. When Boko Haram loots a community, the livelihoods of those effected are destroyed and the group is able to acquire food and funds to continue its operations. The effects are similar when the group loots military targets. The resources of the Nigerian military are depleted while Boko Haram's are replenished. These military looting incidents often arise when the group targets security installations. The group goes after these targets to

demonstrate their operational capacity and to fight the Nigerian State directly.
Finally, not all of the group's bombings are successful and we are able to predict
some of these attempted bombings. Together, all of these events contribute the cocktail of terror Boko Haram has created in northeastern Nigeria. We describe some of
the events we have linked to the various terror attacks below. It also should be noted
that these relationships are not causative.

- When the Nigerian government does not: execute captured militants, close down
  Boko Haram's locations, and forcibly resettle people, the targeting of government officials can be predicted one month ahead of time.
- When the Nigerian government is not receiving military aid, Boko Haram is not
  using unspecified channels to spread its propaganda, and the Nigerian government has not closed the country's borders, we can predict lootings three months
  in advance.
- If the Nigerian government is not receiving military aid and Boko Haram is not
  supporting another NSAG financially, or putting out threats, we can predict the
  group will not target security installations four months in advance.
- If Boko Haram's communications do not address the public or take responsibility
  for an event, and the Nigerian government is not receiving military aid we can
  predict they will not target security installations four months in the future.
- If Boko Haram is using child soldiers, then we can anticipate attempted bombings three months in the future.

Even though the relationships above are not causative, they do provide some
insight into circumstances surrounding some of Boko Haram's attacks. By understanding these circumstances around these attacks, resources and assets can be better allocated to combat them. If looting is expected, security alerts and deployment
of additional personnel could help better guard stores of food, fuel, and weapons. If
security installations are expected to be targeted, garrisons could be put on high
alert or reinforced. Fighting an asymmetrical conflict was shown to be very difficult
throughout the twentieth and twenty-first centuries and the story is no different in
northeast Nigeria. We hope the rules and relationships we have found can make the
task slightly easier.

8.7   Predictive Model Results

Figure 8.3 below shows the performance of our predictive model and predictive
reports during 2019. From January 2019 to the writing of this book, our team has
produced reports containing our predictions on the likelihood of several different
attacks out by Boko Haram. Our predictive model uses over 90 months' worth of
data collected for our research. The dataset is updated at the end of every month to
reflect new occurrences in the Boko Haram conflict. We used 6 classifiers on our
dataset: SVM, KNN, Random Forrest, Gaussian Naïve Bayes, Multinomial Naïve
Bayes, and Logistic Regression. The model aims to predict whether the targeting of

Targeting of Security Installations						
Time Period	1	2	3	4	5	6
Recall	82%	100%	100%	100%	100%	100%
Precision	100%	100%	100%	100%	100%	100%
Accuracy	82%	100%	100%	100%	100%	100%
F1	0.90	1.00	1.00	1.00	1.00	1.00

Fig. 8.2 Targeting of security installations results for our 2019 predictive reports

Looting						
Time Period	1	2	3	4	5	6
Recall	82%	100%	100%	100%	100%	100%
Precision	100%	100%	100%	100%	100%	100%
Accuracy	82%	100%	100%	100%	100%	100%
F1	0.90	1.00	1.00	1.00	1.00	1.00

Fig. 8.3 Looting results for our 2019 predictive reports

Attempted Bombings						
Time Period	1	2	3	4	5	6
Recall	0%	100%	89%	100%	100%	100%
Precision	0%	73%	80%	91%	100%	100%
Accuracy	27%	73%	73%	91%	100%	100%
F1	0.00	0.84	0.84	0.95	1.00	1.00

Fig. 8.4 Attempted bombing results for our 2019 predictive reports

security installations, lootings, or attempted bombings, will, or will not, happen within a given timeframe. For example, if the offset is 2 then the model predicts whether or not an event will occur anytime during the next two months. The Figs. 8.2, 8.3, and 8.4 below depict the results of our predictions compared to the ground truth observed after the predictions we made.

References

Al Jazeera (2018) Boko Haram attacks two military bases in Northeast Nigeria. https://www.aljazeera.com/news/2018/12/boko-haram-attacks-military-bases-northeast-nigeria-181228061504837.html. Accessed 22 Apr 2019

BBC News (2014) Boko Haram spend hours 'killing & looting' in Alagarno attack – BBC News. https://www.youtube.com/watch?v=ELims3HVMVk. Accessed 16 May 2019

CBS News (2016) Boko Haram reportedly on rampage near site of girls' abduction. https://www.cbsnews.com/news/boko-haram-islamic-militants-nigeria-chibok-rampage-schoolgirls-abduction/. Accessed 24 Apr 2019

Fox News (2015) Nigerian Army repels Boko Haram attack on army base. https://www.foxnews. com/world/nigerian-army-repels-boko-haram-attack-on-army-base. Accessed 17 May 2019

Kazeem Y (2018) A Boko Haram attack on a Nigerian army base has left up to 100 soldiers dead. Quartz Africa. https://qz.com/africa/1473661/boko-haram-kill-100-nigerian-soldiers-in-army-base-attack/. Accessed 24 Apr 2019

Kola O (2016) Boko Haram ambush kills 9 in NE Nigeria's Borno State. Anadolu Agency. https:// www.aa.com.tr/en/africa/boko-haram-ambush-kills-9-in-ne-nigerias-borno-state/675414. Accessed 16 May 2019

Mark M (2015) Thousands flee as Boko Haram seizes military base on Nigeria border. The Guardian. https://www.theguardian.com/world/2015/jan/05/boko-haram-key-military-base-nigeria-chad-border. Accessed 24 Apr 2019

NBC News (2014) Boko Haram eyed in attacks on Nigerian Military Base, Mosque. https://www. nbcnews.com/storyline/missing-nigeria-schoolgirls/boko-haram-eyed-attacks-nigerian-military-base-mosque-n148881. Accessed 9 May 2019

News24 (2014) Nigeria: Boko Haram rob bank in looting spree. https://www.news24.com/Africa/ News/Nigeria-Boko-Haram-rob-bank-in-looting-spree-20141105-2. Accessed 9 May 2019

Sahara Reporters (2017) 27 killed, dozens missing after Boko Haram attacks. http://saharareporters.com/2017/08/25/27-killed-dozens-missing-after-bokoharam-attacks. Accessed   24   Apr 2019

The Washington Post (2016) Boko Haram attacks northeast Nigerian city, town, 80 killed. https://www.washingtonpost.com/world/at-least-80-killed-as-boko-haram-attacks-two-nigerian-cities/2015/12/28/08240ae6-ada3-11e5-b820-eea4d64be2a1_story.html?utm_ term=.7eeea7192bfb. Accessed 17 May 2019

Vice News (2015) At least 40 dead as Boko Haram fighters loot and burn villages in Niger. https:// news.vice.com/en_us/article/7xa9mz/at-least-40-dead-as-boko-haram-fighters-loot-and-burn-villages-in-niger. Accessed 24 Apr 2019

Appendices

Appendix A: All TP-Rules

Appendix A contains all of the TP-Rules that have been written up in this book. Table A.1 contains the TP-Rules used to predict attacks, while Table A.2 contains all of the TP-Rules used to predict the absence of attacks.

Table A.1 All of the TP-Rules used to predict attacks

Attack TP-Rules				
Attack type	Offset	Condition A	Condition B	Condition C
Abductions	1	Boko Haram recruits/trains and/or deploys individuals of young age.	Boko Haram is not actively advocating for religious rule.	
Abductions	1	Boko Haram recruits/trains and/or deploys individuals of young age.	Boko Haram is not reported as striving for unspecified political aspirations and objectives.	Boko Haram did not reportedly give military support to another NSAG.
Abductions	1		Security forces have not reportedly executed anyone or the practiced had been abolished.	Boko Haram did not reportedly give military support to another NSAG.
Abductions	2	The Nigerian government closes some Boko Haram locations.	Boko Haram communications did not address the public.	The government was reportedly not elected/postponed/cancelled elections.
Arson	1	Boko Haram is not actively advocating for religious rule.	Security forces have not reportedly executed anyone or the practiced had been abolished.	

(continued)

Table A.1 (continued)

Attack TP-Rules				
Attack type	Offset	Condition A	Condition B	Condition C
Attempted bombings	3	Boko Haram recruits/trains and/or deploys individuals of young age.		
Looting	3	Military aid to the government has been suspended or denied.	The group reportedly did not/denies or stopped utilizing (unspecified) media.	The border has reportedly not been closed or re-opened after closure.
Release	1	Boko Haram recruits/trains and/or deploys individuals of young age.	Boko Haram's members are reportedly not on trial.	
Release	2	Boko Haram recruits/trains and/or deploys individuals of young age.	Boko Haram's members are reportedly not on trial.	Boko Haram reportedly did not use a pseudonym.
Release	3	Boko Haram recruits/trains and/or deploys individuals of young age.		
Release	4	Boko Haram recruits/trains and/or deploys individuals of young age.	The Nigerian government did not engage in sexual violence against civilians.	Boko Haram did not issue messages addressing its aspirations and objectives.
Sexual violence	1	Boko Haram is not actively advocating for religious rule.		
Sexual violence	3	Boko Haram reportedly has members imprisoned by the government.	Boko Haram communications did not address the government.	Boko Haram openly expressed its willingness to negotiate or denied refusing talks.
Sexual violence	2	Boko Haram reportedly has members imprisoned by the government.	Boko Haram reportedly has no foreign members.	Boko Haram did not reportedly give financial support to another NSAG.
Sexual violence	5	Boko Haram reportedly has members imprisoned by the government.	Boko Haram reportedly has no foreign members.	

(continued)

Table A.1 (continued)

Attack TP-Rules

Attack type	Offset	Condition A	Condition B	Condition C
Suicide bombing	1	Boko Haram recruits/trains and/ or deploys individuals of young age.		
Suicide bombing	1	Boko Haram is not actively advocating for religious rule.	Members of Boko Haram were reportedly arrested.	Boko Haram reportedly has no foreign members.
Suicide bombing	1	Boko Haram is not actively advocating for religious rule.	Boko Haram's members are reportedly not on trial.	Negotiations between Boko Haram and the government reportedly did not end with a cease-fire.
Suicide bombing	3	Boko Haram is not actively advocating for religious rule.	Boko Haram did not reportedly communicate their strategy.	
Suicide bombing	3	Boko Haram is not actively advocating for religious rule.	Members of Boko Haram were reportedly arrested.	
Targeting government officials	1	The government did not close down Boko Haram's locations.	Security forces have not reportedly executed anyone or the practiced had been abolished.	Nigerian security forces were not engaged in forcible resettlement activities.
Targeting security installations	4	Military aid to the government has been suspended or denied.	Boko Haram did not issue messages with threats.	Boko Haram did not reportedly give financial support to another NSAG.
Targeting security installations	4	Military aid to the government has been suspended or denied.	Boko Haram did not issue messages with claim of responsibility.	Boko Haram communications did not address the public.

Table A.2 All of the TP-Rules used to predict an absence of attacks

No attack TP-Rules

Attack type	Offset	Condition A	Condition B	Condition C
No Arson	1	Boko Haram was not reportedly recruiting/ training and/or deploying individuals of a young age.	Boko Haram was not designated as an international terrorist organization.	
No Arson	1	The government did not reportedly shut down Boko Haram's offices.	Boko Haram was not designated as an international terrorist organization.	

(continued)

Table A.2 (continued)

No attack TP-Rules				
Attack type	Offset	Condition A	Condition B	Condition C
No Arson	1	Boko Haram was not reportedly recruiting/ training and/or deploying individuals of a young age.	The Nigerian government did not declare a State of Emergency.	
No Arson	1	Boko Haram was not reportedly recruiting/ training and/or deploying individuals of a young age.	International organization(s) reportedly did not seek/deny or revoke prior allegations of human rights violations against Boko Haram.	
No Arson	1	The government did not reportedly shut down Boko Haram's offices.	There was no reported sexual violence by Nigerian security forces.	
No Arson	1	The government did not reportedly shut down Boko Haram's offices.	International organizations did not reportedly freeze assets of members of Boko Haram.	
No Arson	1	Military aid has to the government been suspended or denied.	The Nigerian government did not declare a State of Emergency.	
No sexual violence	1	The government did not reportedly shut down Boko Haram's offices.		
No sexual violence	1	Boko Haram was not reportedly recruiting/ training and/or deploying individuals of a young age.		
No sexual violence	2	Military aid has to the government been suspended or denied.	Foreign state(s)/international institution(s) reportedly did not accuse Boko Haram of human rights abuses.	Boko Haram did not issue messages addressing a justification of violence.
No sexual violence	3	Military aid has to the government been suspended or denied.	Boko Haram reportedly did not/denies or stopped to forcefully recruit people.	No direct negotiations between Boko Haram and the state took place.
No sexual violence	2	Military aid has to the government been suspended or denied.	Boko Haram reportedly did not/denies or stopped to forcefully recruit people.	
No suicide bombing	1	Boko Haram was not reportedly recruiting/ training and/or deploying individuals of a young age.	Boko Haram reportedly did not/denies/or stopped creating YouTube content.	

(continued)

Table A.2 (continued)

No attack TP-Rules

Attack type	Offset	Condition A	Condition B	Condition C
No suicide bombing	1	Boko Haram was not reportedly recruiting/ training and/or deploying individuals of a young age.	No negotiations between Boko Haram/the respective state and/or mediator(s) were planned.	Boko Haram did not reportedly give military support to another NSAG.
No suicide bombing	3	The government did not reportedly shut down Boko Haram's offices	Boko Haram reportedly did not give unspecified support to another NSAG.	
No suicide bombing	4	Boko Haram was not reportedly recruiting/ training and/or deploying individuals of a young age.		

Appendix B: Data Collection

The data we have collected covers the time period from July 2009, to the time of writing of this book. For our dataset we look at whether or not an attack, or event, has occurred during a given month. Months where an event or attack occurred are marked with a 1, while non-occurrence of an event or attack is represented with a 0. Our data was collected using the academic news service Lexus Nexus. For every month of data we collected, we would read the first 300 articles and record the relevant variables. We note that non-English articles were skipped during data collection. Table B.1 shows all of the variables that were recorded – these variables are based on a codebook initially put together by Jana Shakarian for the first author's prior study of Lashkar-e-Taiba, the terrorist group that carried out the infamous 2008 Mumbai attacks.

Table B.1 All of the variables recorded by our data

All recorded variables	
Variable	Explanation
Act_Abd_Abduction General	Did Boko Haram abduct people during the month in question?
Act_Abd_Release General	Did Boko Haram release abducted people during the month in question?
Act_Alt_Assassination	Did Boko Haram assassinate people during the month in question?
Act_Alt_Member Kill	Did members of Boko Haram reportedly kill other members of Boko Haram?
Act_Alt_Sexual Violence	Did Boko Haram commit acts of sexual violence during the month in question?
Act_Armed Clashes – Group's Casualties	During an armed clash, were casualties of Boko Haram reported?
Act_Armed Clashes – Security Forces Casualties	During an armed clash, were casualties of Security Forces reported?
Act_Armed Clashes – Unspecified Casualties	During an armed clash, were unspecified casualties reported?
Act_Ars_Arson General	Did Boko Haram reportedly use arson in an attack?
Act_Attack w/o specifica	Was there an attack without specific details?
Act_Attack_Attempted Attack	Did Boko Haram reportedly attempt an attack?
Act_Attack_Civilian Casualties	Were there civilian casualties?
Act_Attack_Government	Did Boko Haram attack the government?
Act_Attack_Hit & Run	Did Boko Haram commit a hit and run attack?
Act_Attack_Production Site	Did Boko Haram reportedly attack a production site?
Act_Attack_Public Site	Did Boko Haram attack a public site?
Act_Attack_School	Did Boko Haram attack a school?
Act_Bomb_Attempted Bombing	Did Boko Haram attempt a bombing?
Act_Bomb_Bombing General	Did Boko Haram use a non-suicide bombing?
Act_Bomb_Suicide Bomb	Did Boko Haram reportedly commit a suicide bombing?
Act_Bus_Arms Trafficking	Did Boko Haram reportedly engage in arms trafficking?
Act_Bus_Extortion	Did Boko Haram reportedly engage in extortion?
Act_Bus_Human Trafficking	Did Boko Haram reportedly engage in human trafficking?
Act_Bus_Looting	Did Boko Haram reportedly engage in looting?
Act_Bus_Robbery w/o specifica	Did Boko Haram commit unspecified robberies?
Act_Bus_Sexual Exploitation	Did Boko Haram reportedly exploit anyone sexually?
Act_Mnp_Technical Manipulation w/o specifica	Did Boko Haram commit unspecified technical manipulation?
Act_S&C_Public Infrastructure	Did Boko Haram seize public infrastructure?
Act_S&C_Security Force's Structure	Did Boko Haram seize control of a structure belonging to security forces?
Act_S&C_Symbolic Site	Did Boko Haram seize control of a symbolic site?

(continued)

Table B.1 (continued)

All recorded variables

Variable	Explanation
Act_S&C_Territory	Did Boko Haram seize territory?
Act_Z_Civ_Belief	Did Boko Haram target civilians for their beliefs?
Act_Z_Civ_Civilian w/o specifica	Did Boko Haram target civilians without a specific reason?
Act_Z_Civ_Civilians Indiscriminate	Did Boko Haram target civilians indiscriminately?
Act_Z_Civ_Political Orientation	Did Boko Haram target civilians for their political orientations?
Act_Z_Prof_Government Official	Did Boko Haram target government officials?
Act_Z_Prof_School Teacher	Did Boko Haram target teachers?
Act_Z_Prof_Security Force	Did Boko Haram target security forces?
Act_Z_Prof_Production Site Personnel	Did Boko Haram target production site personnel?
Act_Z_Struc_Gov Building	Did Boko Haram target government buildings?
Act_Z_Struc_Security Installation	Did Boko Haram target security installations?
Act_Z_Struc_Production Site	Did Boko Haram target productions sites?
Act_Z_Struc_Public Site	Did Boko Haram target public sites?
Act_Z_Struc_Public Transportation	Did Boko Haram target public transportations?
Act_Z_Struc_Structure w/o specifica	Did Boko Haram target an unspecified structure?
Act_Z_Struc_Symbolic Site	Did Boko Haram target a symbolic site?
Act_t_Election Day	Did Boko Haram attack on an election day?
Com_Add_Addressee w/o specifica	Did Boko Haram communicate without a specific addressee?
Com_Add_Media	Did Boko Haram's communications address the media?
Com_Add_Government	Did Boko Haram's communications address the government?
Com_Add_Public	Did Boko Haram's communications address the public?
Com_Add_Security Forces	Did Boko Haram's communications address security forces?
Com_Add_NSAG	Did Boko Haram's communications address non-state actor groups?
Com_Med_E-mail	Was Boko Haram's communications medium email?
Com_Med_Medium w/o specifica	Was Boko Haram's communications medium not specified?
Com_Med_News Media/Periodicals	Was Boko Haram's communications medium news media?
Com_Med_YouTube	Was Boko Haram's communications medium YouTube?
Com_Mge_Abusive Language	Did Boko Haram's communications contain abusive language?
Com_Mge_Aspirations & Objectives	Did Boko Haram's communications contain aspirations and objectives?

(continued)

Table B.1 (continued)

All recorded variables	
Variable	Explanation
Com_Mge_Call for Recruits	Did Boko Haram's communications contain a call for recruits?
Com_Mge_Call for Violence	Did Boko Haram's communications contain a call for violence?
Com_Mge_Campaign	Did Boko Haram's communications contain a campaign?
Com_Mge_Claim of Responsibility	Did Boko Haram's communications contain a claim of responsibility?
Com_Mge_Declared Enemy	Did Boko Haram's communications establish their enemy?
Com_Mge_Justification of Violence	Did Boko Haram's communications contain a justification of violence?
Com_Mge_Solidarity	Did Boko Haram's communications contain a message of solidarity?
Com_Mge_Strategy	Did Boko Haram's communications contain a strategy?
Com_Mge_Threat	Did Boko Haram's communications contain a threat?
Com_Mge_Hostage	Did Boko Haram's communications mention a hostage?
Env_GovSF_Curfew	Did Nigerian Security Forces enforce a curfew?
Env_GovSF_Desertion	Did Nigerian Security Forces desert?
Env_GovSF_Execution	Did Nigerian Security Forces execute civilians?
Env_GovSF_Forceful Resettlement	Did Nigerian Security Forces forcibly resettle people?
Env_GovSF_Remuneration of Security Forces	Did Nigerian Security Forces get paid?
Env_GovSF_Sexual Violence	Did Nigerian Security Forces commit acts of sexual violence?
Env_GovSF_State of Emergency	Did Nigerian Security Forces enforce a state of emergency?
Env_Gov_Intl_Allegation of Human Rights Abuse	Was Nigerian government accused of human rights abuses?
Env_Gov_Intl_Allegation of War Crime	Was Nigerian government accused of war crimes?
Env_Gov_Intl_Border Closure	Did the Nigerian government close the border?
Env_Gov_Intl_Foreign Aid	Did the Nigerian government receive foreign aid?
Env_Gov_Intl_Sanction, Resolution	Was the Nigerian government sanctioned?
Env_Gov_Intl_Travel Ban	Did the Nigerian government institute a travel ban?
Env_Gov_Intl_Military Aid	Did the Nigerian government receive foreign military aid?
Env_Gov_Leg_Election	Was the Nigerian government elected legitimately?
Group_Basics_A.K.A	Did Boko Haram go by another name?
Group_Basics_Organization Strength	Were there reports on the group's strength?

(continued)

Table B.1 (continued)

All recorded variables

Variable	Explanation
Equ_AA_Anti-Aircraft Weapon	Did Boko Haram reportedly have anti-aircraft weapons?
Equ_AV_Armored Vehicle General	Did Boko Haram reportedly have armored vehicles?
Equ_Art_Artillery General	Did Boko Haram reportedly have artillery?
Equ_B&E_Bombs & Explosives General	Did Boko Haram reportedly have bombs and explosives?
Equ_B&E_IED General	Did Boko Haram reportedly have IEDs?
Equ_F&G_Rifle General	Did Boko Haram reportedly have rifles?
Equ_F&G_Firearms & Guns w/o specifica	Did Boko Haram reportedly have unspecified firearms?
Equ_F&G_Machine Gun General	Did Boko Haram reportedly have machine guns?
Equ_F&G_Small Arms General	Did Boko Haram reportedly have small arms?
Equ_GL_Grenade General	Did Boko Haram reportedly have grenades?
Equ_GL_Rocket Propelled Grenade (RPG)	Did Boko Haram reportedly have RPGs?
Equ_K&S_Knives & Swords General	Did Boko Haram reportedly have knives or swords?
Equ_LV_Logistic Vehicle General	Did Boko Haram reportedly have logistics vehicles?
Equ_Mn_Mine General	Did Boko Haram reportedly have mines?
Equ_PT_Pyrotechnic w/o specifica	Did Boko Haram reportedly have pyrotechnics?
Equ_Rec_Internet Usage	Did Boko Haram reportedly have internet access?
Equ_Rec_Phone	Did Boko Haram reportedly have phones?
Equ_Weapon w/o specifica	Did Boko Haram reportedly have unspecified weapons?
Intra-Org Conflict General	Was there conflict within the Boko Haram organization?
Lead_Att_Academic	Was the leadership of Boko Haram trained or educated?
Lead_Att_Arrest	Was the leadership of Boko Haram arrested?
Lead_Att_Deceased	Was the leadership of Boko Haram killed?
Lead_Att_Fractious	Was the leadership of Boko Haram split/fractured?
Lead_Leg_Election	Was the leadership of Boko Haram elected?
Lead_Leg_External Installation	Was the leadership of Boko Haram installed by an external group?
Memb_Childsoldier	Were members of Boko Haram children?
Memb_Forceful Recruitment	Were members of Boko Haram recruited using force?
Memb_Foreigner	Were members of Boko Haram foreigners?
Memb_Socio-Economical Status	Were members of Boko Haram of a lower socio-economic status?
Group_Rel_Financial Support given	Did Boko Haram give financial support to another NSAG?
Group_Rel_Material Support given	Did Boko Haram give material support to another NSAG?

(continued)

Table B.1 (continued)

All recorded variables	
Variable	Explanation
Group_Rel_Military Support given	Did Boko Haram give military support to another NSAG?
Group_Rel_Unspecified Support given	Did Boko Haram give unspecified support to another NSAG?
A&O_Ego_Profit	Did Boko Haram aim to make money?
A&O_Gov_ID_Non-Religious Rule	Did Boko Haram aim to install a non-religious government?
A&O_Gov_ID_Religious Rule	Did Boko Haram aim to install a religious government?
A&O_Gov_Political Aspirations & Objectives w/o specifica	Did Boko Haram have unspecified political objectives?
A&O_Gov_Political Participation	Did Boko Haram aim to participate politically?
Rel_Alliance General affirmed/formed	Did Boko Haram form an alliance with another NSAG?
Rel_Conflict Non-Violent w/NSAG	Did Boko Haram engage in a non-violent conflict with another NSAG?
Rel_Surrender	Did members of Boko Haram surrender?
Rel_Gov_Amnesty	Did the Nigerian government give amnesty to members of Boko Haram?
Rel_Gov_Arrest	Did the Nigerian government arrest members of Boko Haram?
Rel_Gov_Arrest Release	Did the Nigerian government release arrested members of Boko Haram?
Rel_Gov_Arrest Warrant	Did the Nigerian government issue arrest warrants?
Rel_Gov_Expulsion	Did the Nigerian government expel members of Boko Haram?
Rel_Gov_Presence Close-Down	Did the Nigerian government close down Boko Haram locations?
Rel_Gov_Raid	Did the Nigerian government raid Boko Haram locations?
Rel_Gov_Trial	Did the Nigerian government put members of Boko Haram on trial?
Rel_Intl_Allegation of Human Rights Abuse	Did international actors accuse Boko Haram of human rights abuses?
Rel_Intl_Allegation of War Crime	Did international actors accuse Boko Haram of war crimes?
Rel_Intl_Arrest Extradition	Did international actors arrest and extradite a member of Boko Haram?
Rel_Intl_Designated Terror Organization	Did international actors designate Boko Haram a terror organization?
Rel_Intl_Embargo	Did another country but an embargo on Nigeria?
Rel_Intl_Freeze Asset	Did another country freeze Boko Haram's assets?
Rel_Intl_Travel Ban	Did another country ban travel to or from Nigeria?

(continued)

Table B.1 (continued)

All recorded variables	
Variable	Explanation
Rel_Neg_Current Direct Negotiations	Are negotiations between the Nigerian government and Boko Haram occurring?
Rel_Neg_Current Indirect/3rd Party Talks	Are negotiations between a third party and Boko Haram occurring?
Rel_Neg_End_Ceasefire	Was a ceasefire negotiated?
Rel_Neg_Group Refuse	Did Boko Haram refuse to negotiate?
Rel_Neg_Negotiations Planned	Are negotiations planned between Boko Haram and the Nigerian government?
Rel_Neg_State Refuse	Did the Nigerian government refuse to negotiate with Boko Haram?

Appendix C: Most Frequently Occurring Independent Variables

Appendix C contains a Table C.1, a table of the variables most frequently used by the TP-Rules that we derived from our Boko Haram Data.

Table C.1 A list of the most used variables. All variables used three times or more were included

Most used independent variables in TP-Rules	
Variable	No. of times used
Boko Haram recruits/trains and/or deploys individuals of young age.	9
Boko Haram is not actively advocating for religious rule.	7
Boko Haram reportedly has no foreign members.	4
Boko Haram reportedly has members imprisoned by the government.	3
Boko Haram's members are reportedly not on trial.	3
Military aid to the government has been suspended or denied.	3
Security forces have not reportedly executed anyone or the practiced had been abolished.	3

Appendix D: Sample Report

Boko Haram Action Anticipation Report

Report Date: 31 October 2019
Provided by DSAIL Lab, Dartmouth College

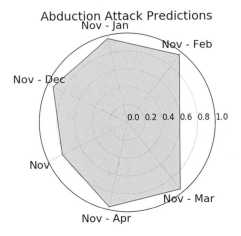

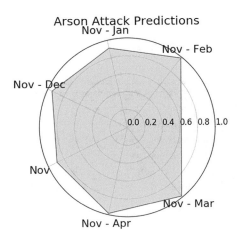

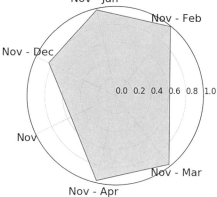

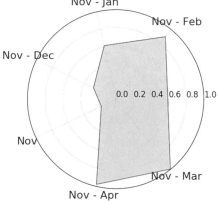

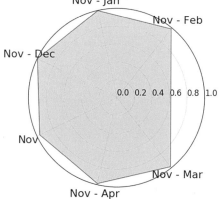

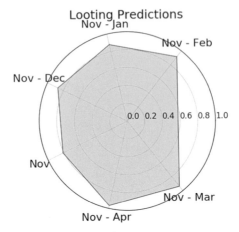

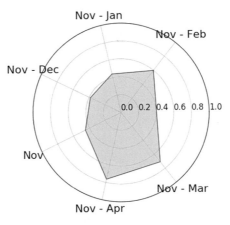

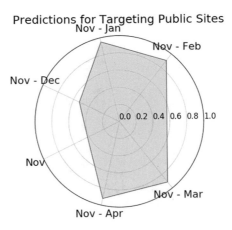

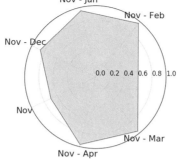

Predictions for Targeting Security Installations

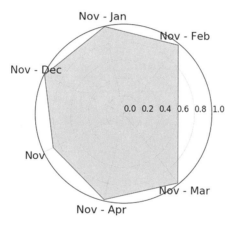

Predictions for the Indiscriminate Targeting of Civilia

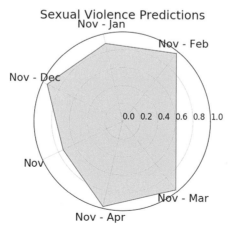

Sexual Violence Predictions

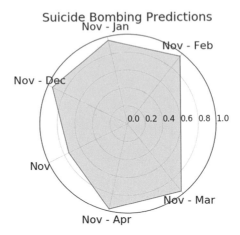

Suicide Bombing Predictions

Section 1.0 – Abduction Attack Predictions
Abductions are very likely and should be expected during all of the time periods that we predict for.

Timeframe	Nov	Nov–Dec	Nov–Jan	Nov–Feb	Nov–Mar	Nov–Apr
Likelihood	0.822	0.931	0.963	0.963	0.957	0.966
Event occurred:	Yes	Yes	Yes	Yes	Yes	Yes

Section 2.0 – Arson Attack Predictions
Arson attacks should be expected during any of the six time periods that we cover.

Timeframe	Nov	Nov–Dec	Nov–Jan	Nov–Feb	Nov–Mar	Nov–Apr
Likelihood	0.889	0.946	0.910	0.987	0.986	0.985
Event occurred:	Yes	Yes	Yes	Yes	Yes	Yes

Section 3.0 – Attempted Bombing Predictions
An attempted bombing is more likely than not during the month of November. The risk of this type of event is elevated during all of the larger time periods.

Timeframe	Nov	Nov–Dec	Nov–Jan	Nov–Feb	Nov–Mar	Nov–Apr
Likelihood	0.557	0.839	0.983	0.994	0.978	0.975
Event occurred:	No	Yes	Yes	Yes	Yes	Yes

Section 4.0 – Bombing Attack Predictions
A non-suicide bombing is very unlikely to occur during November, and the time period examining November through December. During the November through

January time period it is slightly more likely than not that a non-suicide bombing will occur. For the remaining three time periods a non-suicide bombing is very likely.

Timeframe	Nov	Nov–Dec	Nov–Jan	Nov–Feb	Nov–Mar	Nov–Apr
Likelihood	0.186	0.285	0.614	0.897	0.997	0.977
Event occurred:	Yes	Yes	Yes	Yes	Yes	Yes

Section 5.0 – Civilian Casualty Predictions
Civilian casualties are extremely likely during all of the time periods that we examine.

Timeframe	Nov	Nov–Dec	Nov–Jan	Nov–Feb	Nov–Mar	Nov–Apr
Likelihood	0.966	0.997	0.997	0.986	0.985	0.987
Event occurred:	Yes	Yes	Yes	Yes	Yes	Yes

Section 6.0 – Looting Predictions
Looting is extremely likely to occur in November, as well as all of the other time periods.

Timeframe	Nov	Nov–Dec	Nov–Jan	Nov–Feb	Nov–Mar	Nov–Apr
Likelihood	0.812	0.868	0.879	0.917	0.938	0.967
Event occurred:	Yes	Yes	Yes	Yes	Yes	Yes

Section 7.0 – Predictions for Targeting Public Sites
During November an attack on a public site should not be anticipated. However, during the time period examining November through December the risk of this type of targeting increases and becomes slightly more likely than not. During the remaining four time periods this type of targeting becomes very likely.

Timeframe	Nov	Nov–Dec	Nov–Jan	Nov–Feb	Nov–Mar	Nov–Apr
Likelihood	0.421	0.525	0.951	0.904	0.911	0.921
Event occurred:	Yes	Yes	Yes	Yes	Yes	Yes

Section 8.0 – Predictions for Targeting Security Installations
It is likely that a security installation will be targeted during any of the six time periods that we predict for.

Timeframe	Nov	Nov–Dec	Nov–Jan	Nov–Feb	Nov–Mar	Nov–Apr
Likelihood	0.697	0.866	0.952	0.969	0.964	0.970

Timeframe	Nov	Nov–Dec	Nov–Jan	Nov–Feb	Nov–Mar	Nov–Apr
Event occurred:	Yes	Yes	Yes	Yes	Yes	Yes

Section 9.0 – Predictions for Targeting Civilians for Their Beliefs
It is unlikely that civilians will be targeted for their beliefs during the three smallest time periods. However, during the largest three time periods targeting of this type become likely.

Timeframe	Nov	Nov–Dec	Nov–Jan	Nov–Feb	Nov–Mar	Nov–Apr
Likelihood	0.451	0.386	0.442	0.598	0.710	0.763
Event occurred:	Yes	Yes	Yes	Yes	Yes	Yes

Section 10.0 – Predictions for the Indiscriminate Targeting of Civilians
The indiscriminate targeting of civilians should be expected during any time period we predict for.

Timeframe	Nov	Nov–Dec	Nov–Jan	Nov–Feb	Nov–Mar	Nov–Apr
Likelihood	0.881	0.993	0.996	0.983	0.987	0.986
Event occurred:	Yes	Yes	Yes	Yes	Yes	Yes

Section 11.0 – Sexual Violence Predictions
The risk of sexual violence is high during all of the time periods we predict for.

Timeframe	Nov	Nov–Dec	Nov–Jan	Nov–Feb	Nov–Mar	Nov–Apr
Likelihood	0.742	0.945	0.892	0.975	0.976	0.980
Event occurred:	Yes	Yes	Yes	Yes	Yes	Yes

Section 12.0 – Suicide Bombing Predictions
Suicide bombings should be expected during any of the time periods we predict for.

Timeframe	Nov	Nov–Dec	Nov–Jan	Nov–Feb	Nov–Mar	Nov–Apr
Likelihood	0.744	0.945	0.956	0.962	0.967	0.977
Event occurred:	No	Yes	Yes	Yes	Yes	Yes

Section 13.0 – How to Use This Report
This report aims to provide some insight into what Boko Haram may do in the coming months. In the preceding sections we presented how likely we think certain events are to happen within a specified number of months. It is important to recognize that these predictions are not meant to precisely predict every individual attack

made by the group. Our predictions should be used to gain an understanding of what types of attacks, or what events, are most likely to unfold in a future time frame. For example, if we predict that there is a 74% chance of suicide bombings one month in the future, we expect that a suicide bombing will occur at some point within the next month. In this report we presented some of the events we can predict well. In this report we shared our predictions for Sexual Violence, Bombings, Suicide Bombings, Arson, Civilian Casualties, Attempted Bombing, Attacks on Civilians for Their Beliefs, Indiscriminate Civilian Killings, Looting, Attacks of Public Sites, and Attacks on Security Installations. We made predictions for the likelihood of these events happening within the next month, two months, three months, four months, five months, and six months.

For each of the time periods we predicted for, we have an annotation to indicate whether or not the event has occurred. Three are three types of annotations. "No" indicates that the event did not occur within the given time period. "Not Yet" indicates that the time period is not over and the even has not occurred yet. Finally, "Yes" indicates that the event did occur within the time period. It should be noted that all reports are generated with "Not Yet" being the default annotation. Annotations are adjusted to reflect actual events at the end of each time period.

Section 14.0 – About DSAIL Lab

The Dartmouth Security and AI Lab (DSAIL) develops data-driven computation models to help address the world's security problems, covering cyber-security, counter-terrorism, and defense/intelligence related issues. DSAIL also performs fundamental research in artificial intelligence, machine learning and knowledge discovery, and spatio-temporal-probabilistic logic models. Contact us at dsail-info@cs.dartmouth.edu and on the web at www.TEMP.com.

The Boko Haram Analysis Toolkit was generously funded by the Office of Naval Research under Grant Number N000141612739.

Printed in the United States
by Baker & Taylor Publisher Services